ABOUT ISLAND PRESS

Island Press is the only nonprofit organization in the United States whose principal purpose is the publication of books on environmental issues and natural resource management. We provide solutions-oriented information to professionals, public officials, business and community leaders, and concerned citizens who are shaping responses to environmental problems.

In 1994, Island Press celebrated its tenth anniversary as the leading provider of timely and practical books that take a multidisciplinary approach to critical environmental concerns. Our growing list of titles reflects our commitment to bringing the best of an expanding body of literature to the environmental community throughout North America and the world.

Support for Island Press is provided by The Geraldine R. Dodge Foundation, The Energy Foundation, The Ford Foundation, The George Gund Foundation, William and Flora Hewlett Foundation, The John D. and Catherine T. MacArthur Foundation, The Andrew W. Mellon Foundation, The Joyce Mertz-Gilmore Foundation, The New-Land Foundation, The Pew Charitable Trusts, The Rockefeller Brothers Fund, The Tides Foundation, Turner Foundation, Inc., The Rockefeller Philanthropic Collaborative, Inc., and individual donors.

REINVENTING
NATURE?

REINVENTING NATURE?

RESPONSES TO POSTMODERN DECONSTRUCTION

EDITED BY

MICHAEL E. SOULÉ AND GARY LEASE

Illustrations by Alan Gussow

ISLAND PRESS

WASHINGTON, D.C. • COVELO, CALIFORNIA

Art at chapter openings by Alan Gussow. Information regarding these pieces appears at the back of the book.

Reinventing nature? : responses to postmodern deconstruction / edited
  by Michael E. Soulé and Gary Lease ; illustrations by Alan Gussow.
     p. cm.
    Includes bibliographical references and index.
    ISBN 1-55963-310-7 (cloth : acid-free paper). — ISBN 1-55963-311-5 (paper : acid-free paper)
    1. Environmental sciences—Philosophy.   2. Philosophy of nature.
3. Human ecology.   4. Deconstruction.   5. Postmodernism.   I. Soulé,
Michael E.   II. Lease, Gary, 1940–
GE40.R45   1995
304.2 — dc20   94-22631
                                             CIP

Printed on recycled, acid-free paper

♻

Manufactured in the United States of America
10 9 8 7 6 5 4 3 2 1

CONTENTS

The boundary between the world and human beings is under fire. On the one hand nature is personified; on the other hand the idea that nature needs protection from humankind's onslaught begs the definition of the boundary and turns our attention to contesting constructions of nature and to competition among human groups for access to resources and power. Whose story (narrative, paradigm, construction) will prevail? Deconstruction insists that we must not ignore these cultural questions, even in the formerly exclusive provinces of science and conservation. Reverberations of past issues are sensed in debates over what is "out there," what is nature, and the locus of the human species vis-à-vis the divine. The Western tradition has not found the answers. Postmodern answers, to date, have ignored certain actors and obscured certain questions—for example, the issue of conceptual constructions of nature versus the role of human beings in the physical construction of ecosystems. The answers will affect the lives of many.

The postmodern constructionist view is that all texts, reports, narratives are but descriptions—focused chatter about an unknowable external world, psychobabble, webs of words that serve as ammunition in struggles over who dominates whom. But Derrida, Lyotard, and other deconstructionists have about them the smell of the coffeehouse, a world of ironic, patronizing remoteness in which the search for generality and truth would be an embarrassment. Moreover, somehow justified by the deconstruction of nature are the theme parks, malls, and other virtual simulations of originals that create a world easier to control, a world where imagination is the only real landscape and where denial replaces even disengagement and relativism. The loss of contact with nature, a biophilic deprivation, must lead to pathology. But other animal species, because they have no words to confuse themselves, are not so deluded.

For most of human time, reality/nature were divine and one thing. The Greek thinker Thales stands for the rupture of this spiritual plenum. By the sixteenth century, nature was seen as the recalcitrant power against which people had to struggle, then vanquish—leading, in the late twentieth century, to the end of nature. Now the fragmented, anthropogenic wilderness cannot be left unattended lest it deteriorate even faster. Modernity paints this dismal scenario, a melancholy acquiescence to decline; but this view rests on a false dichotomy: natural vs. artificial, independent vs. managed. An alternative to ubiquitous artificiality is the admission of degrees of "reality." The criteria are genuineness, seriousness, and commanding presence. Thus the substitute for the dualism of natural and artificial is a new continuum: reality—hyperreality. And even if nature (reality) is to some extent a human invention, it still can be eloquent and inspiring and still can invigorate the notion of excellence. A general guideline: to save or restore a wild area's commanding presence and to guard its coherence with its environment and tradition.

The deconstructionist paradigm, if accepted broadly, would not only threaten the privileged role of science as a source of truth about reality. It would also destroy environmentalism, since the environment is just a "social construction." Short of dismissing the institution of science, not to mention all of "reality," we might adopt a *constrained* program of deconstruction using the concepts of interactivity and positionality. This program may create a common ground between adherents of the strong program of deconstruction, on the one hand, and traditional objectivists (for example, Western scientists) on the other. For example, human beings develop their models of reality with their sensory/cognitive apparatus, "the cusp," each person being uniquely embodied and positioned. But not all representations of reality are equally acceptable because certain constraints, such as consistency across cultures, can falsify representations. (Thus this "constrained constructivism" does not mean that all realities are equally valid, as does the strong program.) The notions of interactivity and positionality enliven the stakes in contesting for the integrity of the environment. Those in power, therefore, should consider marginal points of view, including those of other species.

Attitudes about nature and the environment change. But contrary to the belief of many contemporary postmodern historians, whose excessive relativism may distort reality, change may not be the most important metaphysical principle. Still, disorderly change is the fashion of the day. Just as in ecology, where the Victorian paradigm of stability, equilibrium, and order has been superseded by a paradigm of disturbance and disorder, the contemporary historian's view of human society rejects the notions of normality, equilibrium, progress, and all value judgments; it is fixated on disorder. As Marx said: "All that is solid melts into air, all that is holy is profaned." Because modern historicism leads either to cynicism or to banality, it could be described as a degenerate worldview. A less extreme interpretation of contemporary history and ecology might stress two principles: one is social and biological interdependence; the other is successful adaptation to situation and place by human groups and species. Change is not a good in itself. But preserving a diversity of change, not freezing nature, ought to stand high in our system of values.

Does conservation of wilderness imply excluding residents who practice traditional forms of human subsistence? The debate over this issue is relevant to the question of the past human role in the "construction" of native ecological systems. What is original, untrammeled nature—primitive America? Is it pre-Columbian, implying that Native Americans walked softly and lived in harmony? Or did they and their ancestors deforest large areas, cause the mass extinction of mammals, and change the landscape everywhere by burning? It is clear now that Native Americans practiced extensive and intensive land management, though this evidence was often invisible to the European settlers who arrived after epidemics had erased it. In any case, the polarized debate about aboriginal impacts obscures the complexity and diversity of old cultures in North America and ignores cultural adaptation and change. Such local, cultural knowledge of nature in indigenous groups is rapidly being lost because the mass media expose Native American children to pan-Indian culture and a generic electronic nature.

Even though nature evokes common emotional and intellectual structures in humans (evidence for a shared understanding), cultures are heterogeneous in how they *value* nature. The scholarly depiction of Western (Judeo-Christian) perspectives as oppressive and exploitive, and Eastern views as unitary, respectful, and harmonious, may not be relevant or even accurate. Survey results comparing Americans and Japanese indicate that Japanese are inclined to emphasize control over nature, to emphasize the presumed essence of natural phenomena, especially in contrived or symbolic representations, and to be less knowledgeable about natural history and ecology. Americans are far more likely to express concern about ethical treatment of animals and to support environmental protection. Attempts to attribute the aloofness of Japanese to Western intellectual colonization seem simplistic, for Eastern attitudes toward nature have always been idealized and passive. Though the seeds of a conservation ethic can be found in both cultural traditions, they are not the same seeds. The deconstructionist notion that all cultural perspectives of nature possess equal value is both biologically misguided and socially dangerous.

National parks (and wilderness in general) are by default the sites where the values of solitude, wildness, and otherness reside. Yet the baseline criteria for original, natural, or pristine states still elude managers. And public pressure for aesthetics and recreation militates against the achievement of biodiversity protection (saving all the parts)—including the control of exotic species and the restoration of native species and associations—and forces abandonment of the principle of "organic autonomy" and the vision of wildness. Management for biodiversity in national parks is incompatible with management for wildness because it requires heroic and intrusive interventions, depriving visitors of the subjective experience of wildness. If a third objective is added—reconstructing cultural landscapes, a current management fad—any sense of wildness recedes further yet. One can only hope that research provides a way out.

Humanity entertains manifold representations of living nature—from quite pagan/spiritual views to the more utilitarian (Judeo-Christian) and scientific conceptions displayed on television documentaries. Myths about ecological equilibrium, homeostasis (including Gaia), nature's pristine/profane dualism, and human population underlie some of these representations, as do the postmodern social myths of Western moral inferiority, historicism, and cultural relativity. The overt siege of nature, entraining a major extinction episode, is sometimes defended by various ideologies—the social siege and its postmodern premises. Yet there is evidence against the more rigid constructionist/historicist positions, and it can be shown how postmodern views have been employed by the Wise Use movement, the Animal Rights movement, and the Social Ecology and Justice movement to justify further exploitation of wildlands. Some positive elements of the postmodern, humanist agenda—a focus on power inequities, bias, and the myths that maintain them—may help balance the harm to living nature by the politics of deconstruction.

A NOTE FROM THE EDITORS

The following are terms with which our readers may not be familiar. Those terms lacking a definition or explanation can be found in any dictionary.

construction Any interpretation of reality; a model, theory, opinion, belief, or value that claims to represent reality.

constructionism See "deconstructionism."

deconstructionism A theory current in contemporary literary criticism, philosophy, and history, which claims that no text has an originally fixed meaning and therefore is re-created in a new form with each reading. The work of the critic (philosopher, historian) thus consists of breaking down every text into the infinite number of meanings it is capable of expressing, none of which is definitive. Social critics often use the term "deconstruct" to mean an analysis of premises and biases, sometimes to justify normative relativism.

meta (-discipline, -narrative) A field or area, or an account of such a field or area, that deals critically with other fields or areas of study previously held to be independent.

modernism A movement of cultural expression and critical study, originating at the end of the nineteenth century, that denied authoritative permanency to traditional conceptions of reality and thus sought to multiply any culture's conceptions of its defining realities.

postmodernism A cultural and critical movement that took form after World War II and which advanced modernism to its final conclusion, namely that no authoritative and definitive expression or conception of reality is possible.

semiotics Both a theory and a method of understanding language, and thus texts, as systems of signs and symbols. Every language statement is produced according to a previous "code" of such signs and symbols; to understand the meaning of an expression, one must first master the code that preceded its utterance.

social constructivism Action based on the ideological persuasion that reality is the product of social interactions and dynamics.

Preface

This multidisciplinary volume is a response to certain radical forms of "postmodern deconstructionism" that question the concepts of nature and wilderness, sometimes in order to justify further exploitive tinkering with what little remains of wildness. An eminent European colleague warned us not to publish this work because, he said, the deconstructionists feed on controversy. Perhaps he is right, but we feel that the threats to nature are now so grave that the prudent course is to directly challenge some of the rhetoric that justifies further degradation of wildlands for the sake of economic development.

This book, then, is about a clash of intellectual fashions. The so-called deconstructionist view—the one that the authors of this book modify, analyze, or critique to some degree—asserts that all we can ever perceive about the world are shadows, and that we can never escape our particular biases and fixed historical–cultural positions. Moreover, some in the deconstructionist movement boldly assert that the natural world as described by scientists and conservationists, if it exists, is a human artifact produced by our economic activities, and as such is grist for further material reshaping.

It is ironic that similar notions have also been adopted by representatives of both the capitalist right (the "wise use" movement) and the animal rights movement. The capitalists ask society for a license to maximize short-term profits from the aggressive exploitation of natural resources, while some of the animal rights activists seek to minimize the physical pain of feral or introduced animals such as goats, sheep, pigs, house cats, and foxes at the expense of native biodiversity.

The opposing view, defended to varying degrees by the authors, assumes that the world, including its living components, really does exist apart from humanity's perceptions and beliefs about it. Most of the authors agree that we can gain dependable, scientific knowledge about this independent, natural world, in spite of differences among us in class, culture, gender, and historical perspective. Taking the offensive, some chapters critique the deconstructionist argument that nature isn't natural because aboriginal human beings altered—physically constructed—contemporary ecosystems with the use of fire and other manipulations. Finally, some of the authors worry that the absence of contact with nature during childhood and adolescence has significant social consequences, not least of which is the dissolving of human diversity into one global, economic, consumerist monoculture.

Thus the reader will discover a diversity of viewpoints that cannot be classified as simply "Western" as opposed to "multicultural," or "pro-wilderness" as opposed to humanist. While some of the authors simply reject the entire deconstructionist project, many if not all are sympathetic with the goals of emancipatory social movements.

It would be extraordinary if this diverse collection of contributors—ethnobiologists, historians, a literary critic, a philosopher, an artist, a sociologist, and two zoologists—were to agree on anything. But they do. They agree that certain contemporary forms of intellectual and social relativism can be just as destructive to nature as bulldozers and chain saws. Here, then, is a nascent interdisciplinary synthesis on the nature of nature—one that is neither left nor right, positivist nor relativist, but one that rejects some popular fads in academia. We hope that this book will encourage those who wish to engage in a constructive dialogue that aims to protect nature as well as serve humanity.

The following institutions and people provided assistance with this book and the conference that preceded it. The University of California Humanities Research Institute and its Reinventing Nature project directed by Mark Rose, with the local guidance of Jim Clifford, provided the key funding that made the conference and this book possible. Additional funding was provided by the Humanities Division of the University of California, Santa Cruz. David Orr, David Ehrenfeld, and Vivian Sobchack provided advice early on. Ray Dasmann, Ed Grumbine, Carolyn Merchant, Todd Newberry, and Gary Snyder commented on the proceedings. Mark Rose, Jim Clifford, Daniel Press, Virginia Sandoval, and others helped facilitate discussions at the conference. We also wish to thank Barbara Dean

for her advice and wisdom. Conference logistics were facilitated by
Maggie Collins, the staff of the Environmental Studies Boards, and
Joy McKinney.

Michael E. Soulé
Gary Lease

REINVENTING NATURE?

CHAPTER ONE

INTRODUCTION: NATURE UNDER FIRE

GARY LEASE

"There is no form of prose more difficult to understand and more tedious to read," insists Nobel laureate Francis Crick, "than the average scientific paper."[1] He, of course, should know. But the natural sciences are certainly not alone in producing thick, even obscure, language. Much of contemporary writing from the humanities can certainly be opaque and difficult, particularly in the areas of theoretical, feminist, neohistorical, deconstructive, and ultimately critical thought. Witness Donna Haraway, eminent feminist theorist and historian of science, who noted in a recent talk at the University of California, Santa Cruz, that the nonhuman world is dialogic—indeed, a coproductive participant in human social relationships. This is not as it should be, emphasizes Haraway, but it is the way things are. For her the question is not one of difference between the human and the nonhuman worlds but rather the nature of nonnature, which she views as a product of techno-science. Haraway finds the most blatant example in the patenting of "invented" animals; here we do not have simply animals as property, an old practice, but animals as human invention.[2] From her point of view the contemporary notion of nature takes the form of a contest over the politicization of nature—that is, the objectification of nature and its distinction from the political and the social.[3]

What can Haraway mean? Perhaps public opinion can help. In a Christmas Day (1993) poll released by the *Los Angeles Times*, 47 percent of those interviewed thought that animals "are just like humans in all important ways."[4] The boundaries, in other words, between the human species and other animal species are increasingly permeable in our highly self-reflective Western cultures. In fact, the boundaries between the world and humans, after much careful construction, seem to be under fire. Nature is not only personified— "it's not nice to fool Mother Nature," sang one margarine commercial not so long ago—but often carefully reduced. The world of human production has frequently operated outside of nature. Cities, transportation, recreation: all those myriad inventions of human culture have been viewed as inimical to the wild and (humanly) uncontrolled state of the world, a condition akin to a New Eden. Haraway wants to focus attention on the fact that even the attempt to rescue the nonhuman from the human, to salvage nature from the onslaught of modern, technologized humanity, is itself a human construction. What, then, is nature? Where is it to be found? Who controls it? Should it be controlled at all? What is the relationship of humans to that nature: are they in it, out of it, or somewhere in between? There is a war over nature in progress and nature itself is in the middle—caught in a crossfire of competing interests.

Such a contest is not over empty prizes. Indeed, it is nothing less than the human struggle for access to reality. And for humans, access to something is what grants control: nine-tenths of the law is possession, goes the old saw, and such clichés have power precisely because they so often are the case. Without access to the understanding of something, one is powerless over it. If one does have that access, however, and is able to control the communication over a phenomenon, then one controls how it is to be understood. A contest over what is allowed to represent reality—and that is what intelligible access is all about—is a struggle over that reality itself. This is the heart of our age's modernism, the process by which we establish what "counts" as reality.

Seen against this background, all our narratives—our many stories about nature and ourselves, whether "scientific" or not—are striving for such representation. In effect, we work hard to establish reality. In this hurly-burly world of contestation, the results are always more exclusive than inclusive. We are constantly deciding what belongs and what does not. In both the sciences and the humanities, pedigrees are the goal—in our stories we invariably exclude far more than we include. And that is also the problem: nature and human are both places (embodiment) and narratives, sites of contestation which can never be resolved. How do we map and present such places? How is such a place visited or experienced? In other words, what are the mechanics of contestation, the strategies and tactics (Clausewitz) of representation? Or: how can one reduce narrative access to possession, and how can one access the access?

In our postmodern, post-Marxist world, class struggles no longer have anything to do with "truth," with "right" and "wrong," but rather only with the most profound level of ideological battles; in the last analysis it is always a question of life and death, of pure survival. Such contests never result in victory, in completion, in closure. We will not "get the story right," regardless of the tendency of some scientists to proclaim final triumph, and despite the hubris of some historians who announce that their stories have finally portrayed events "as they actually were" (Ranke). Our many representations of nature and human are, in other words, always and ultimately failures. The reason for such failure is, of course, not far to seek: nature and human are not self-revealing, even to a self-reflective species such as the human one. We and our world may well be real, but intelligible access to that reality is constructed and produced and ultimately incomplete. Yet we need to form judgments about these constructions; otherwise we would not be able to tell our stories. This, in turn, underlines the role of *power* in the contestation over what gets to count in any ruling narrative, and who gets to tell it. It is this struggle that is basically the story of modernism. And modernism is not fun.

"I don't want to just hear about revolutions. All we see or hear is revolutions. I'm sick of them," said Hemingway's female protagonist. But he saw it differently. "They're beautiful," he wrote. "Really. For quite a while. Then they go bad." Nevertheless, he studied them. The result: "They were all very different but there were some things you could co-ordinate." Only under certain conditions, of course; for one thing you have to have enough material. In fact: "You need an awful lot of past performances. It's very hard to get anything true on anything you haven't seen yourself because the ones that fail have such a bad press and the winners always lie so."[5]

This is the story being replayed today in the ever accelerating disintegration of the natural sciences and the humanities in their traditional forms at universities and research institutes across the globe. In the humanities, such stalwarts as literature and history are shifting under the impact of deconstruction, new historicisms, and the rise of new disciplinary configurations such as women's studies and feminist theory, a revitalized (revisionist?) notion of world history, and metadisciplines galore. Deconstruction, for example, in the field of literary criticism has led to an expansion of concerns that takes all human expression to be "text" and thus the proper target of literature. In fact, literature is too restrictive; for all intents and purposes, what goes on under the rubric of "textual" studies is cultural studies: the investigation of all human production. Or take history (Please! cries Henny Youngman, dean of comedy in any contemporary university), where the impossibility of ever accumulating all the sources, witnesses, and evidence to any moment in the past has transformed the writing of history into publicly acknowledged fiction (Schama). Elsewhere the natural sciences seem to be splitting faster than zygotes, creating right and left new groupings and directions. In response, Stanford University hosted this year a major conference to probe the world of thought "beyond dualism," seeking a realignment of the sciences and humanities. It is risky business indeed to speak these days of the "natural" sciences and the "humanities" in the monolithic tones common just ten years ago. For one thing, the practitioners no longer share anything like a common understanding of what it is they are about; for another, the transformation of all human expression and production, including scientific experimentation and knowledge, into "texts" to be deconstructed according to their ranking on the scales of power and control has washed out any previous lines of difference. A revolution in the making?

When Michael Soulé, chair of Environmental Studies at the University of California (Santa Cruz), and I, in my capacity as dean of the humanities at the same institution, first heard of a large-scale, three-year project on Reinventing Nature being planned by the University of California Humanities Research Institute (Irvine), we were immediately intrigued. Taking its inspiration from a recent publication by Donna Haraway,[6] the project sought to promote a series of regional conferences throughout California, ultimately to culminate in a residential research team in Irvine during the first half of 1994. The overall initiative was understood as a foray into the world of environmental change, seeking to understand the dynamics unleashed by two major trends in today's world. The first trend is the recognition that the forces of cultural construction play a much greater role in forming our understanding of nature than has been admitted; the second trend is the acknowledgment of the still strong defense of nature as a realm that is autonomous and valuable in its own right. Conferences in Berkeley (March 1992), San Diego (November 1992), and Davis (September 1993) sought to deal with the problems of narrative and image, with the arts, and with the notion of "wilderness." What struck both Soulé and me from the very beginning was the fact that none of these conferences was designed to address specifically the dialogue between the worlds of the natural sciences and the humanities. In planning the conference that has led to this book, we sought to fill this gap.

Our initial premise was that the current reanalysis of nature, ecology, and wilderness in contemporary discourse, including the social sciences, has important cultural implications.[7] We knew that such a statement might well shock many scientists and technocrats—partly for "turf" reasons but also because they may be unaware of the dialectics occurring in other disciplines. Scientists and conservationists, when confronted with the concepts of invention and deconstruction, as applied to nature, sometimes fail to appreciate the degree to which their own concepts of nature are culturally determined. Despite the problems of communication across this "two-culture" gap, we believed that all the parties, including historians, sociologists, and biologists, have much to learn from each other and that cross-disciplinary contacts can promote teaching and research collaborations, the advance of knowledge, and the development of more effective conservation policies.[8]

From the start we were convinced that the inherited notions of "nature" as well as the distinctions between the invention of nature as cultural construction and the invention of specific biophysical contents should be the main points of departure for any such dialogue. Of course we are not the first to pose such questions. The place of humanity in nature—or, more precisely, the relationship of the human species to the rest of reality—has been a central problem in all historical cultures, but most particularly in the Western tradition that has produced both our contemporary "sciences" and the "humanities." This question is, of course, partly epistemological: Is nature "out there" or do we create it? Cultural historians and analysts frequently distinguish between "nature" as other and humans, who stand somehow outside of nature.[9] Such a move is as old as the question itself. Certainly the early Greek Sophists depended on a distinction between the "original" or the essential, on the one hand, and that which is artificial, acquired, or accidental, on the other. The contempt of the Cynics for conventional custom was determined by this distinction as well. Augustine, one of the most influential early Christian thinkers, betrayed in his *Confessions* both a Manichean (dualist) background and a Christianized Neoplatonism by considering "nature" to be the original act of divine creation; humanity, though created, is destined to achieve separation from nature and unity with God. From Pseudo-Dionysus through the early medieval Platonists to the reworked Aristotelianism of Aquinus, there emerged a concept of "natural" law, or a divinely ordained "order" to all reality. Such an order established a framework within which the human species and the physical world were interrelated in subordination to the creative will of God.[10] In other words, for a large and formative part of our Western tradition, nature has been a theological given, linked always to a divine transcendence and defined in contrast to that deity.

It was British philosophy of the seventeenth and eighteenth centuries, however, that brought a systematic rejection of this theological scheme of understanding the world. Locke, in particular, insisted on the distinction between that which is "made," or relations, and that which is "real," or experienced. This view led inevitably to an opposition between reason and nature, a position which Kant in his idealism effectively exploited. And while the early Greek thinkers often restricted the term "nature" to the world of the physical and material, it was Spinoza who finally drew the consequences of a natural law stance and identified nature with God.[11] Thus Western thought had culminated in an impasse regarding nature. Was it the material world of experience, experiences that could be shared, repeated, and tested; or was it the ineffable, invisible, and transcendent world of divine origins, available only to acts of faith? The answer of the nineteenth century was clear. Any notion of a functioning "natural" law in Western tradition declined markedly, a direct result of the disappearance of a theologically grounded base both to public life and to scientific pursuit. After wrestling with the notion of nature for well over two thousand years, Western tradition had come up dry: neither an identification of the human species with nature nor a strict dichotomy between the two proved ultimately successful.

In our so-called postmodern world, however, this struggle has been revived. On the one hand, the effort continues to remove the human species from nature. This can lead, in one scenario, to an overvaluation of the human role in creating or forming nature, thus producing a paradoxical identification of human and natural. On the other hand, there is an insistence on an overdetermined independence of humans from world/nature, resulting in separation. In both cases, though, the relationship between human cultural production and the biophysical world may be dangerously skewed. One of the objectives of our Santa Cruz conference was to shift the focus of such discussions away from such polar extremes and into more fruitful, if complex, intellectual discourse.

The need for such a redirected conversation is made even more obvious by the fact that all too often barriers akin to cultural boundaries have been erected between humanistic emphasis on the role of human conception in establishing what is "natural" and scientific insistence on nature as "given." Major aspects of these "two-culture" problems with respect to "nature/ecology/wilderness" have rarely if ever been exposed to multidisciplinary dialogue; as a result, many confusing issues and semantic debates continue to hamper the discovery of the basic issues. We saw earlier, for example, that construction of nature, espoused so pointedly by Haraway, can be interpreted in at least two ways: one is the cultural *context* in which nature is understood; the other involves the actual *content* and *structure* of that nature.

The former issue, context, is conceptual and perceptual—it is about our view of nature more than about nature's actual composition, chemically, physically, and biologically, which, arguably, is largely an empirical question. While the social construction of nature is hardly an issue any more among social scientists, the intellectual and practical significance of the social construction is an open question. Clearly there are important differences between "Western" and "non-Western" cultural constructions of nature, for example, but these can be exaggerated (see Kellert in Chapter 7). As Soulé explains (Chapter 9), the plant taxonomies of aboriginal societies are virtually always the same in structure as those of modern, scientific ones—both are hierarchical, consisting of nested sets of ex-

clusive categories—and aboriginal taxonomies typically recognize
the same entities as species.[12] The fundamental question both Soulé
and I wanted to raise at the conference is whether perceptions and
conceptions of nature (landscapes, animal/human relations, ecological processes) differ enough between cultures to affect the way these
cultures would wish to maintain or manage nature in the remnants of
remaining habitat. In other words, do these differences have policy
implications? (See Graber in Chapter 8; Soulé in Chapter 9.)

The impact of humans on nature's *content* and *structure*, however, is
another matter. Most observers agree that aboriginal groups have
over long periods of occupancy altered contemporary ecosystems in
a concrete way by the use of fire, selective harvesting, selective plantings, and similar economic activities. Indeed, social critics have argued, based on ecological evidence, that South and Central American
Indians have "constructed" the rainforest.[13] (See Nabhan in Chapter
6; Graber in Chapter 8; Soulé in Chapter 9.) This means, of course,
that humans and nature exist in a dialectical relationship, each imag-
ining the other. Such a relationship is marked by history—a history
of contingency.[14] (See Shepard in Chapter 2; Worster in Chapter 5.)
This symbiotic connection is best represented in the oscillation between two poles of explanation: constructionism (the insistence on
the human making of what counts as nature) on the one hand and es-
sentialism (the emphasis on the independent, unchanging character
of the nonhuman world) on the other.

This volume of essays is dedicated to exploring and illuminating
this tense but unavoidable relationship. As we observed earlier, much
older, influential models of the world viewed nature as an inclusive
"creation" (Cicero, Lucretius, Augustine, Aquinas). But the struggle
over "artifact" (ultimately, technological inventions) as a separate
human creation led inevitably to the exclusion of humanity from nature. This move was finalized in the seventeenth century, when the
divine transcendent was relocated to human subjectivity. In essence,
divinity as transcendence was assumed by humanity (Descartes). At
least for Western tradition, this shift effectively placed humans outside of nature—leading to the explosion of psychology as a separate,
"natural," and experimental science in the nineteenth century, to the
Freudian turn to the human subject and its formative origins in human relationships, and thus to the issues that are the chief objects of
attention in the essays that follow:

- The fundamental differences between the academic cultures in regard to their views of nature and related concepts
- The degree to which scientific, historical, and ideological paradigms may produce these differences
- The degree to which ecological "facts" and principles are dependent on contemporary constructions of nature and, in turn, how these "facts" are changing (reinventing) nature in both its cultural and physical meanings
- How the current revolution in ecological theory, including the importance of nonlinear and chaotic dynamics, bears on these issues
- Whether perceptions and conceptions of nature (wilderness, landscapes, animal/human relations, ecological processes) differ enough between cultures to affect the ways these cultures would wish to maintain or manage nature in the remnants of remaining habitat (as in national parks) and how these contrasting views actually affect public policy and the management of nature and wildlands
- The implications of contemporary claims that modern ideas of wilderness and conservation are "Western," often elitist, neocolonial concepts
- The ethical questions raised by a constructionist perspective— including the role of humans in organic evolution—and whether these questions are likely to be manifest in public policy issues such as the release of bioengineered organisms and the rights of animals
- The possible influences of the current nature debate on the practices and policies of wildland management, hunting, and the protection of biodiversity

The conference's intense, two-day exchange, marked by a vibrant conversation between the speakers and the participants, led, in turn, to the following problematics—still visible in the essays in this volume—which are an invitation to the reader for participation in an extended discussion:

- Who precisely defines "nature" —that is, who is allowed to say what counts as nature and why? This is, of course, the question of the "reinvention" of nature or, more precisely, the question of power and privilege: not only do we wish to discover who reinvents nature, but who invented it to begin with? What are the tools (language, culture, nation-state, science, academia, and so forth) by which such inventions are sustained in power, and how does one supplant or overthrow them? In other words, this is the question of revolution and defense.

- The scope or extent of inclusion/exclusion—that is, who belongs and why? More to the point, this is about nature's content; it touches not only on what is allowed to belong to nature, and why, but also on who is allowed to participate in nature and what happens to those parts and beings excluded from nature. In other words, this is the question of survival and destruction.

- The shared character of the conversation—that is, can we agree on what and who is in or out? Here we are concerned with the issue of how we include others, both human and nonhuman, in the determination of what is allowed to count as nature. With whom do we reach such conclusions? In other words, this is the question of definition and imposition.

- The actual construction of nature—that is, how is nature constituted as a "unity"? The final resolution of dialogue between the sciences and humanities about nature is to be found in the content of that nature: to what degree are we able to speak together about common experiences and realities rather than talking past each other's heads in the dark? In other words, this is the question of intelligibility and understanding.

Do not think that these questions do not matter. They do. As Norman Maclean points out in his utterly moving account of the Mann Gulch fire and the smokejumpers who died in it: "If you have lived a life that has thrown you in contact many times with nature, you have already discovered that sometimes you can deal with nature only by allowing it to push back what until now you and others thought were its edges."

The result was that the Mann Gulch fire passed "beyond issues and settlements into a world of pictures—perhaps more exactly into thoughts that pictorialize and feel and cannot reduce themselves to numbers. These are pictures made largely by us, the amateur artists who are always making pictures inside our heads (that spring from our hearts)."

And what is the final picture? As Maclean tells us how those men died, running from the fire, caught by it, dying as the flames finally overhauled them but rising a last time to flee again, he notes that "the evidence, then, is that at the very end beyond thought and beyond fear and beyond even self-compassion and divine bewilderment there remains some firm intention to continue doing forever and ever what we last hoped to do on earth. By this final act they had come about as close as body and spirit can to establishing a unity of themselves with earth, fire, and perhaps the sky."[15]

Such unity is perhaps quixotic. Given our markedly varied points of departure, ranging from history to biology, from ecology to public policy, the conference's discussions, and the essays in this book, could not have been otherwise. Language and individual experience, it turns out, are not unproblematic and self-evident. These two primary sources of human understanding are not illusionary, but neither do they represent direct and immediate access to reality. At best they are mediational, and thus always limited.

Even more troublesome is the consequence that language and the appeal to experience do not permit untrammeled sharing. How, then, are they to function as the sources of policy? In the end we must admit, together with the contributors to this volume, that the identity crises besetting both the sciences and the humanities at the end of the twentieth century are crises of norms and values. Those versions of nature, those "pictures," as Maclean called them, that were once so self-evident and stable, that were shared so widely among us, are in disintegration. To wrestle with what nature is, and how humans are part of it, and yet are not its entirety, stands in the finest tradition of both the humanities and the sciences; striving to "see," or devise new pictures because the old ones no longer function.

Notes

1. In his latest study, *The Astonishing Hypothesis: The Scientific Search for the Soul* (New York: Scribner's, 1994), p. xiii.

2. Haraway was referring to the "oncomouse" from Dupont/Harvard. The talk was delivered as a colloquium address to the History of Consciousness program in February 1994.

3. See Haraway's forthcoming book on nature at the second millennium.

4. As quoted in *Safari Times*, February 1994, p. 18.

5. Ernest Hemingway, *The Green Hills of Africa* (New York: Scribner's, 1935), pp. 192–193.

6. Donna J. Haraway, *Simians, Cyborgs, and Women: The Reinvention of Nature* (New York: Routledge, 1991).

7. For examples of this reanalysis see W. McKibben, *The End of Nature* (New York: Random House, 1989); M. Oelschlaeger, *The Idea of Wilderness* (New Haven: Yale University Press, 1991); G. Snyder, *The Practice of the Wild* (Berkeley: North Point, 1990). For examples in the social sciences see S. Hecht and A. Cockburn, *The Fate of the Forest: Developers, Destroyers and Defenders of the Amazon* (New York: Harper Perennial, 1990); Haraway, *Simians, Cyborgs, and Women*; D. E. Goodman and M. R. Redclift, *Refashioning Nature: Food, Ecology, and Culture* (London and New York: Routledge, 1991).

8. M. E. Soulé, "Conservation: Tactics for a Constant Crisis," *Science* 253 (1991): 744–750; G. Orians, *Ecological Knowledge and Environmental Problem-Solving: Concepts and Case Studies* (Washington, D.C.: National Academy Press, 1986).

9. See, for example, Noel Perrin, "Forever Virgin: The American View of America," in *On Nature: Nature, Landscape, and Natural History*, edited by Daniel Halpern (San Francisco: North Point, 1987), pp. 13–22.

10. See Thomas Aquinas, *Summa Theologiae*, I, II, q. 93, art. 1, where the "eternal" law is identified as divine wisdom, or the essential order of being with which all creation is imbued. But in I, II, q. 94, art. 3, he also views "natural" law as containing "everything to which a man is inclined according to his nature."

11. See his *Ethics*, props. 14 and 15.

12. C. D. Brown, *Language and Living Things: Uniformities in Folk Classification and Naming* (New Brunswick, N.J.: Rutgers University Press, 1984); B. Berlin, P. H. Raven, and D. Breedlove, *Principles of Tzeltal Plant Classification: An Introduction to the Botanical Ethnography*

of a Mayan-Speaking People of Highland Chiapas (New York: Academic Press, 1974).

13. See Hecht and Cockburn, *The Fate of the Forest*.

14. See Stephen Jay Gould's provocative statements on science as history in *Wonderful Life* (New York: Norton, 1989), p. 51.

15. Norman Maclean, *Young Men and Fire* (Chicago: University of Chicago Press, 1992), pp. 277, 289, 300.

CHAPTER TWO

VIRTUALLY HUNTING REALITY IN THE FORESTS OF SIMULACRA

PAUL SHEPARD

It would be difficult to argue with the assertions that our representations of the world are always "interpretations," that concepts shape
our perceptions, that the human organism is its own shuttered window. Here I wish to explore the conclusion that reality is therefore invented by words, much the way Benjamin Whorf claimed seventy
years ago that colors are a consequence of their differentiation by
names. In its current form this idea further argues that such inventions are motivated by a struggle for power and that there is no Grand
Truth beyond texts which are "allusions to the conceivable which
cannot be presented."[1] I am also interested in the relationship of this
inaccessibility of reality to "virtual reality."

According to the postmodern view, what most of us think of as
simulations are only focused chatter about an unknowable external
world. Even though they use living materials, the practitioners of
"restoration ecology" may now find themselves, in this deconstructionist view, in the same boat with the museum curators making di-
oramas or habitat groups, who claim to be making artificial repro-
ductions of the past or present world—and who are, therefore,

merely engaged unconsciously in a sort of paranoid babble, lost in the vapors of their own imaginations without a compass or a satellite.

A 1973 essay in *Science* asks: "What's Wrong with Plastic Trees?"[2] The essay hardly deals with the question it poses; primarily it is the last gasp of a spurious "argument" between "preservation" and "conservation." But the question is important. More recently the curator of the Devonian Botanical Garden, writing in *The Futurist*, points to fabricated lawns and polyester Christmas trees which do not wilt in polluted environments or have to be watered, giving his blessing to the wonderful world of "artificial nature."[3] For years I have bedeviled my students with the issue of the validity of surrogates for living organisms and the enigma of the mindset in which this kind of ambiguity arises, wherein the nature of authenticity and the authenticity of nature are riddled with qualification.[4] For a time I thought the issue could be clarified by examining the megaphysiology of exchange by natural trees with their environment, health-giving to the soil, air, and other organisms, as contrasted to poisoning the surroundings by the industrial pollution from making plastic trees. Now I see that such comparisons no longer matter, since what I said to them, my own psychobabble, is itself the subject.

Reality—You Can't Get There from Here

Looking behind the facades of grafted signification was the intent in 1949 of Marshall McLuhan's brilliant book *The Mechanical Bride*, which dissolved the rhetoric of magazine advertisements to expose the tacit messages concealed in the hypocrisy, presumptions, and deceit of the corporate purveyors of consumerism and our own lust to be seduced.[5] According to the current literary fashion, however, McLuhan himself can now be deconstructed and his own agenda shown to be just another level of presumption and the struggle for power.

If McLuhan still lived in the ocean of positivist naïveté, David Lowenthal, the geographer, appeared on the new, dry shore of equivocal reality only thirteen years later. In an essay called "Is Wilderness Paradise Enow?" he argues that the substantive reality of wilderness exists solely in the romantic ideas of it. Even its inhabitants are fictions, the noble savage having been the first victim of the "new criticism." Even the buffalo, Lowenthal says, "is only a congeries of feelings."[6] The buffalo is not some *thing* among other things—the cowbirds, Indians, pioneers, and you and I. It exists only as the feelings that arise from our respective descriptions.

Lowenthal's argument is reminiscent of the psychological conception of life as being locked inside a series of boxes and therefore precluding us from knowing anything but our own internal pulsations: coded patterns of electrochemical stimuli. At the level of cerebral axons, nuclei, and their glial cohorts the assertion that my world is more real than yours seems ridiculous. And suddenly we are back in Psychology 101, where we realize that all contact with "reality" is translated to the brain as neural drumbeats, nothing more, and in Psych 102, where the instructor titillates the class with the sensational observation that no old tree crashes down on an island if there is no ear to hear it. A thrill runs through the class, who will go on to Literature 101 to learn that the other impulses arise in words, a barrier behind which is a vast, unknowable enigma—or, perhaps, nothing at all.

Many of us—including myself—may think of a photograph as the visual evidence of a past reality, so that certain events may then be recalled or better understood. But we are now confronted with the assertion that there is nothing "in" a picture but light and dark patches or bits of pigment—that the events to which we supposed such photographs refer are not themselves in the blobs. If something actually occurred it cannot be known. The result, in the case of photographs of starving people, is insensitivity to human suffering. This callous aesthetic is the object of Susan Sontag's anathema, identifying her as grounded—like McLuhan—rather than detached with the postmodern solipsists.[7]

Reflected light from actual events, focused through a lens onto a chemically sensitized plate, inscribing images that can be transferred by means of more light, lens, and photochemicals onto paper, does not, however, neutralize the "subject matter" even if it is a century old. The assay of such a photograph on formal grounds is a form of aesthetic distancing. This surreality of pictures which denies the terms of their origin reminds us that, according to the fashion, such a picture is a text, an impenetrable facade, whose truth is hidden.

Likewise, Hal Foster observes that recent abstract painting is only about abstract painting. Paintings are no more than the simulation of modes of abstraction, he says, made "as if to demonstrate that they are no longer critically reflexive or historically necessary forms with direct access to unconscious truths or a transcendental realm beyond the world—that they are simply styles among others."[8] As painting becomes a sign of painting, the simulacra become images without resemblances except to other images. Like the events which occasioned the photographs, the original configurations to which abstractions refer no longer have currency. Reality has dissolved in a connoisseurship of structural principles. A twentieth-century doubt has interposed itself between us and the world. "We have paid a high enough price for the nostalgia of the whole and the one," says Lyotard, and so allusion of any kind is suspect.

In my view this denial of a prior event is an example of what Alfred North Whitehead calls "misplaced concreteness." Paradoxically, the postmodern rejection of Enlightenment positivism has about it a grander sweep of presumption than the metaphysics of being and truth that it rejects. There is an armchair or coffeehouse smell about it. Lyotard and his fellows have about them no glimmer of earth, of leaves or soil. They seem to live entirely in a made rather than a grown world; to think that "making" language is analogous to making plastic trees, to be always on the edge of supposing that the words for things are more real than the things they stand for.[9] Reacting against the abuses of modernism, they assert that life consists of a struggle for verbal authority just as their predecessors in the eighteenth century knew that life was a social struggle for status or a technological war against nature. Misconstruing the dynamics of language, they are the final spokesmen of a world of forms as opposed to process, for whom existence is a mix of an infinite number of possible variations making up the linguistic elements of a "text."

Under all narrative we find merely more layers of intent until we realize with Derrida, Rorty, Lacan, Lyotard, and critics of visual arts that our role as human organisms is to replace the world with webs of words, sounds, and signs that refer only to other such constructions. Intellectuals seem caught up in the dizzy spectacle and brilliant subjectivity of a kind of deconstructionist fireworks in which origins and truth have become meaningless. Nothing can be traced further than the semiotic in which everything is trapped. The chain of relationships that orders a functional fish market, the cycle of the tree's growth, breath, decay, and death, the underlying physiological connections that link people in communities and organisms in ecosystems or in temporal continuity—all are subordinate to the arguments for or against their existence. The text—the only reality—is comparable only to other texts. Nothing is true, says Michel Foucault, except "regimes of truth and power." It is not that simulacra are good or bad replicas—indeed, they are not replicas at all; they are all there is! In a recent essay, Richard Lee characterizes this attitude as a "cool detachment and ironic distanciation," an eruption of cynicism caused by our daily bombardment of media fantasy, assaults on the "real," and consequent debasement of the currency of reality.[10] But I think it is not, as he concludes, simply a final relativism. There is no room even for relative truth in a nihilistic ecology.

The deconstructionist points with glee to the hidden motivations in these "falsifications" of a past and perhaps inadvertently opens the door to the reconfiguration of places as the setting of entertainment and consumption. This posture is not only a Sartrean game or artistic denial. It spreads throughout the ordinary world, where pictorial, electronic, and holographic creations, architectural facades for ethnic, economic, and historical systems, pets as signifiers of the animal kingdom, and arranged news are the floating reality that constitute our experience. We seem to be engaged in demonstrating the inaccessibility of reality.

In the past half century we have invented alternative worlds that give physical expression to the denial of disaster. Following the lead and iconography of *The National Geographic* magazine with its bluebird landscapes, and then the architecture of Disneyland happiness, a thousand Old Waterfronts, Frontier Towns, Victorian Streets, *documentary work is a counter to this*

Nineteenth-Century Mining Communities, Ethnic Villages, and Wildlife Parks have appeared. One now travels not only in space while sitting still but "back" to a time that never was. As fast as the relics of the past, whether old-growth forests or downtown Santa Fe, are demolished they are reincarnated in idealized form.

As the outer edges of cities expand, the centers are left in shambles, the habitation of the poor, or they are transformed into corporate wastelands which administer distant desolation as if by magic. To console the middle-class inhabitants and tourists, a spuriously appropriated history and cityscape replace the lost center with "Oldtown." In 1879 Thomas Sargeant Perry, having looked at Ludwig Friedlander's book on romanticism, wrote: "In the complexity of civilization we have grown accustomed to finding whatever we please in the landscape, and read in it what we have in our own hearts."[11] An example would be the "monumental architecture" of the rocky bluffs of the North Platte River as reported by the emigrants on the Oregon Trail in the 1840s.[12] Were Perry present today he could say that we make "whatever we please" out there and then announce it as found.

Michael Sorkin speaks of this architectural "game of grafted signification . . . [and] urbanism inflected by applique" and the "caricaturing of places." Theme parks succeed the random decline of the city with their nongeography, their surveillance systems, the simulation of public space, the programmed uniformity sold as diversity. Condensed, they become the mall.[13] It is as though a junta of deconstructionist body-snatchers had invaded the skins of the planners, architects, and tour businessmen who are selling fantasy as history, creating a million Disneylands and ever-bigger "events" for television along with electronic playsuits and simulated places in three-dimensional "virtual reality." Apart from the rarefied discourse and intimidating intellectualism of the French philosophes, their streetwise equivalents are already at work turning everyday life into a Universal Studios tour. It is not just that here and there in malls are cafés representing the different national cuisines but that the referent does not exist. Who cares about authenticity with respect to an imaginary origin?

The point at which the architectural fantasists and virtual realists intersect with intellectual postmoderns and deconstructionists is in the shared belief that a world beyond our control is so terrifying that we can—indeed, must—believe only in the landscapes of our imagination.

Amid the erosion of true relicts from our past, can we not turn to the museum? What are the custodians of the portable physical relicts doing? According to Kevin Walsh, museums now show that all trails lead to ourselves, create displays equating change with progress, and reprogram the past not so much as unlike ourselves but as trajectory toward the present.[14] The effect is like those representations of biological evolution with humankind at the top instead of the tip of one of many branches. In effect the museum dispenses with the past in the guise of its simulation, "sequestering the past from those to whom it belongs." Its contents are "no longer contingent upon our experiences in the world" but become a patchwork or bricolage "contributing to historical amnesia." Roots in this sense are not the sustaining and original structure but something adventitious, like banyan tree "suckers" dropped from the ends of its limbs. "Generations to come," Walsh predicts, "will inherit a heritage of heritage—an environment of past plu-perfects which will ensure the death of the past."

Not long ago I was in the Zoological Museum at St. Petersburg, Russia, an institution which has not reorganized its exhibits according to the new fashion of diorama art—those "habitat groups," simulated swamps, seacoasts, prairies, or woodlands, each with its typical association of plants and animals against a background designed to give the illusion of space. There in Russia, among the great, old-fashioned glass cases with their stuffed animals in family groups, with no effort at naturalistic surroundings, I felt a rare pleasure. I realized that the individual animal's beauty and identity remain our principal source of satisfaction. When all members of the cat family, or the woodpeckers, are placed together, instead of feeling that I am being asked to pretend that I am looking through a window at a natural scene, I am free to compare closely related forms. Instead of an ersatz view I have the undiluted joy of those comparisons that constitute and rehabilitate the cognitive processes of identification. Nor does he museum display in St. Petersburg insidiously invade my thoughts as a replacement for vanishing woodlands and swamps by substituting an aesthetic image for noetics, the voyeur's super-real for the actuality.

New Dress for the Fear of Nature

Plastic trees? They are more than a practical simulation. They are the message that the trees which they represent are themselves but surfaces. Their principal defect is that one can still recognize plastic, but it is only a matter of time and technology until they achieve virtual reality, indistinguishable from the older retinal and tactile sensa. They are becoming *acceptable configurations*. No doubt we can invent electronic hats and suits into which we may put our heads or crawl, which will reduce the need even for an ersatz mock-up like the diorama. These gadgets trick our nervous systems somewhat in the way certain substances can fool our body chemistry—as, for example, the body fails to discriminate radioactive strontium 90 from calcium. (Strontium was part of the downwind fallout of atomic bomb testing which entered the soil from the sky, the grass from the soil, the cows from the grass, the milk from the cows, children from the milk, and finally their growing bones, where it caused cancer.) As the art of simulacrum becomes more convincing, its fallout enters our bodies and heads with unknown consequences. As the postmodern high fashion of deconstruction declares that the text—or bits cobbled into a picture—is all there is, all identity and taxonomy cease to be keys to relations, to origins, or to essentials, all of which become mere phantoms. As Richard Lee says of the search for origins in anthropology, any serious quest for evolutionary antecedents, social, linguistic, or cultural ur-forms, has become simply an embarrassment.

But is this really new or is it a continuation of an old, antinatural position that David Ehrenfeld has called "the arrogance of humanism"?[15] Mainstream Western philosophy, together with the Renaissance liberation of Art as a separate domain and its Neoclassical thesis of human eminence, were like successive cultural wedges driven between humans and nature, hyperboles of separateness, autonomy, and control. It may be time, as the voices of deconstruction say, for much of this ideological accretion to be pulled down. But as down-pullers trapped in the ideology of Art as High Culture, of nothing beyond words, they can find nothing beneath "text." Life is indistinguishable from a video game, one of the alternatives to the physical wasteland that the Enlightenment produced around us. As the tourists flock to their pseudo history villages and fantasylands, the cynics take refuge from overwhelming problems by announcing all lands to be illusory. Deconstructionist postmodernism rationalizes the final step away from connection: beyond relativism to denial. It seems more like the capstone to an old story than a revolutionary perspective.

Alternatively, the genuinely innovative direction of our time is not the final surrender to the anomie of meaninglessness or the escape to fantasylands but in the opposite direction—toward affirmation and continuity with something beyond representation. The new humanism is not really radical. As Charlene Spretnak says: "The ecologizing of consciousness is far more radical than ideologues and strategists of the existing political forms . . . seem to have realized."[16]

Life—or Its Absence—on the *Enterprise*

The question about plastic trees assumes that nature is mainly of interest as spectacle. The tree is no longer in process with the rest of its organic and inorganic surroundings: it is a form—like the images in old photographs. As the plastic trees are made to appear more like natural trees, they lose their value as a replacement and cause us to surrender perception of all plants to the abstract eye. Place and function are exhausted in their appearance. The philosophy of disengagement certifies whatever meanings we attach to these treelike forms—and to trees themselves. The vacuum of essential meaning implies that there really is no meaning. A highbrow wrecking crew confirms this from their own observations of reality—that is, of conflicting texts.

There is a certain bizarre consequence of all this which narrows our attention from a larger, interspecies whole to a kind of bedrock ethnos. For example, in each episode of the television series "Star Trek" we are given a rhetorical log date and place, but in fact the starship *Enterprise* is neither here nor there, now or then. Its stated mission to "contact other civilizations and peoples" drives it frantically nowhere at "warp speed" (speed that transcends our ordinary sense of transit), its "contacts" with other beings so abbreviated that the substance of each story rests finally on the play of interpersonal dynamics, like that which energizes most drama from Shakespeare to soap opera.

This dynamics is *essentially* primate—their own physicality is the only connecting thread to organic life left to the fictional crew and its vicarious companions, the audience. For all of its fancy hardware, software, and bombastic rhetoric, the story depends on projections of primatoid foundations, a shifting equilibrium like any healthy baboon social dynamic—the swirl of intimidation, greed, affection, dominance, status, and groupthink that make the social primates look like caricatures of an overwrought humanity.

The story implies that we may regress through an ecological floor—the lack of place and time and the nonhuman continuum—to farcical substitutes and a saving anthropoid grace. Should such a vehicle as the *Enterprise* ever come into existence, I doubt that their primatehood will save its occupants from the madness of deprivation—the absence of sky, earth, seasons, nonhuman life, and, finally, their own identities as individuals and species, all necessary to our life as organisms.

In the later series, "Star Trek: The Next Generation," the spacecraft has a special room for recreation, the "holo-deck," in which computers re-create the holographic ambience and inhabitants of any place and any time. The difference between this deck and H. G. Wells's time-travel machine is that the *Enterprise* crew is intellectually autistic, knowing that other places and times are their own inventions. The holo-decks that we now create everywhere, from Disneyworld to Old Town Mainstreets, may relieve us briefly from the desperate situations we face on the Flight Deck as we struggle in "the material and psychic waste accumulating everywhere in the wake of what some of us still call 'progress.'" The raucous cries of post-Pleistocene apes on the *Enterprise* are outside the ken of the ship's cyborg second mate, just as the technophiles on earth will witness without comprehension the psychopathology of High Culture and literary dissociation, with its "giddy, regressive carnival of desire," rhetoric of excess, "orgies of subjectivity, randomness," and "willful playing of games."[17]

What, then, is the final reply to the subjective and aesthetic dandyism of our time? Given our immersion in text, who can claim to know reality?

As for "truth," "origins," or "essentials" beyond the "metanarratives," the naturalist has a peculiar advantage—by attending to species who have no words and no text other than context and yet among whom there is an unspoken consensus about the contingency of life and real substructures. A million species constantly make "assumptions" in their body language, indicating a common ground and the validity of their responses. A thousand million pairs of eyes, antennas, and other sense organs are fixed on something beyond themselves that sustains their being, in a relationship that works. To argue that because we interpose talk or pictures between us and this shared immanence, and that it therefore is meaningless, contradicts the testimony of life itself. The nonhuman realm, acting as if in common knowledge of a shared quiddity, of unlike but congruent representations, tests its reality billions of times every hour. It is the same world in which we ourselves live, experiencing it as process, structures, and meanings, interacting with the same events that the plants and other animals do.

Notes

1. Jean-Francois Lyotard, *The Post-Modern Condition*, quoted in Jane Flax, *Thinking Fragments: Psychoanalysis, Feminism and Postmodernism in the Contemporary West* (Berkeley: University of California Press, 1990).

2. Martin H. Krieger, "What's Wrong with Plastic Trees?" *Science* 179 (1973):446–454.

3. Roger Vick, "Artificial Nature, the Synthetic Landscape of the Future," *Futurist* (July–August 1989).

4. See Umberto Eco, "The Original and the Copy," in Francisco J. Varella and Jean-Pierre Dupuy Crea (eds.), *Understanding Origins* (Boston: Kluwer, 1992).

5. Marshall McLuhan, *The Mechanical Bride: Folklore of Industrial Man* (New York: Vanguard Press, 1951).

6. David Lowenthal, "Is Wilderness Paradise Enow?" *Columbia University Forum* (Spring 1964). See also my reply, "The Wilderness as Nature," *Atlantic Naturalist* (January–March 1965).

7. Susan Sontag, *On Photography* (New York: Dell, 1977).

8. Hal Foster, "Signs Taken for Wonders," *Art in America* (June 1986).

9. I used to think of this attitude as the "Marjorie Nicolson Syndrome." It was from her book *Mountain Gloom and Mountain Glory* that I first got the sense there were those who seemed to think the test of nature was whether it lived up to the literary descriptions of it. Clearly, Nicolson neither invented nor bears the full responsibility for this peculiar notion.

10. Richard B. Lee, "Art, Science, or Politics? The Crisis in Hunter-Gatherer Studies," *American Anthropologist* 94 (1993):31–54.

11. Thomas Sargeant Perry, "Mountains in Literature," *Atlantic Monthly* 44 (September 1879):302.

12. An account appears in Paul Shepard, "The American West," in *Man in the Landscape* (College Station: Texas A&M University Press, 1991).

13. Michael Sorkin, "See You in Disneyland," in Michael Sorkin (ed.), *Variations on a Theme Park* (New York: Hill & Wang, 1992).

14. Kevin Walsh, *The Representation of the Past: Museums and Heritage in the Post-Modern World* (London: Routledge, 1992).

15. David Ehrenfeld, *The Arrogance of Humanism* (New York: Oxford University Press, 1981).

16. Charlene Spretnak, *States of Grace* (San Francisco: Harper, 1991), p. 229.

17. Klaus Poenicke, "The Invisible Hand," in Gunter H. Lenz and Kurt L. Shell (eds.), *Crisis of Modernity* (Boulder: Westview Press, 1986).

CHAPTER THREE

THE NATURE OF REALITY AND THE REALITY OF NATURE

ALBERT BORGMANN

To speak of "the nature of reality and the reality of nature" is to play on words. More than that, however, it is to have a helpful guide because the words rightly suggest that the ways we consider and transform the structure of reality have a bearing on the power and presence of nature. And the title further implies—or, at any rate, I will draw this inference from it—that the kind of power and presence nature has is intimately connected with the moral quality of human conduct. Accordingly I will trace the connections between the nature of reality, the reality of nature, and human conduct in today's radically new, deeply disquieting, but finally hopeful constellation.

Just to provide a backdrop, I begin with the original human condition: the hunting and gathering culture that prevailed for some hundreds of thousands of years. In those days, as far as we know, both the nature of reality and the reality of nature were divine. The world was full of divinities: it was a spiritual plenum. Nature, reality, and divinity were one. The human attitude that corresponded to this unified world was one of piety. Not that people were invariably pious. But it was understood that punishment and misery would be visited upon the impious sooner or later. Here on this continent we are at least vicariously within hailing distance of our original condition thanks to the heritage of the Native Americans. In the world of the Blackfeet, for instance, spirits were everywhere and spoke to humans in dreams and visions, through plants, animals, and the powers of the seasons.[1]

To the extent, however, that we have accepted Western Civilization, we are heirs of a different culture. It first began to stir in Ancient Greece. Much of its literature bespeaks the unity of nature (*physis*), reality (*ta onta*), and the divine (*to theion*), a unity the Greeks called beauty and order (*kosmos*). But at the beginning of the sixth century BC, we witness the intellectual departure that eventually led to the fragmentation of the cosmos. Thales was the first to displace piety toward the order of reality with curiosity about the composition of reality. Not that the people of the hunting and gathering cultures lacked curiosity. Yet for them curiosity, rather than being an inclusive attitude, was enfolded in piety. Thales, in any event, suggested that all there is consists of water in various stages of condensation and rarefaction. The world, he maintained, consists of a certain kind of stuff. At least for the philosophers, it was no longer a spiritual plenum. Reality, nature, and divinity no longer could be one. Nature, in due course, became a nonhuman region *of and within* reality, divinity a region *above and beyond* reality. Curiosity about the composition of reality led to the limitation of nature and to the transcendence of the divine.

But the development to that stage took its course slowly and subterraneously. The Jewish, Islamic, and Christian Middle Ages achieved a vision of a divinely instituted order of reality. The natural and the supernatural were distinguished and reconciled through creation and revelation. Nature was created, divinity was not. Divinity needed to be revealed, nature did not. Faith was the dominant human attitude. But its rule was that of a constitutional monarch. Though faith ruled, it had to share some of its authority with reason.

When the medieval order began to collapse toward the end of the fifteenth century, philosophers once more raised the question of the composition of reality. This time they did so not out of curiosity as much as from a reconstructive zeal—from a desire to determine the ultimate components of reality so that, knowing their properties and possibilities, they would be able to construct a world of human liberty and prosperity. In light of this project, nature was seen as the recalcitrant power that kept humans in the bondage of disease and poverty and had to be forced into yielding its secrets and treasures.[2]

The domination of nature gathered momentum as natural philosophy developed into physics and physics into chemistry. The increasingly sophisticated and powerful insights into the composition of reality led to the eradication of diseases, the multiplication of agricultural yields, the discovery of novel kinds of energy and materials, and to a life that for the citizens of the advanced industrial countries is in definite ways far freer and richer than it was in premodern circumstances.

If the modern project of controlling reality and dominating nature was to succeed, it had to culminate in a pervasive artificiality and the end of nature. Bill McKibben has given definition and currency to this view. The epochal event for him is the advent of a new atmosphere.[3] Until recently we could, when mourning the ravages inflicted on nature by human recklessness, take consolation from the immutable and unreachable forces of the heavens, wind and rain, heat and cold. But with the arrival of acid rain, the greenhouse effect, and the ozone hole, the sky and its works have been deeply affected by the work of humans. We have rendered sun and rain injurious and upset the pattern of the weather and the seasons.

McKibben is surely right in saying that the change of atmosphere marks a closure in the history of nature and humanity. But how significant is the closure? And what are we to learn from it? We can begin to respond by considering the Northern Rockies. They are as yet untouched by acid rain, the greenhouse effect, and the ozone hole, and millions of acres have been set aside as wilderness areas. The federal Wilderness Act of 1964 defines such areas as follows:

> A wilderness, in contrast with those areas where man and his own works dominate the landscape, is hereby recognized as an area where the earth and its community of life are untrammeled by man, where man himself is a visitor who does not remain. An area of wilderness is further defined to mean in this Act an area of undeveloped Federal land retaining its primeval character and influence, without permanent improvements or human habitation, which is protected and managed so as to preserve its natural conditions and which (1) generally appears to have been affected primarily by the forces of nature, with the imprint of man's work substantially unnoticeable; (2) has outstanding opportunities for solitude or a primitive and unconfined type of recreation; (3) has at least five thousand acres of land or is of

sufficient size as to make practicable its preservation and use in an unimpaired condition; and (4) may also contain ecological, geological,
or other features of scientific, educational, scenic, or historical value.[4]

And yet the imprint of human work on the fauna and flora of the
wilderness is inevitable and ubiquitous. Fire suppression has allowed
Douglas firs to invade the open stands of mature ponderosa pine. Formerly, periodic low-intensity fires would clear out the understory
while leaving the pines with their bare lower trunks and heavy bark
unscathed. Soon many of these parklike groves with their golden columns and open canopies will be a thing of the past. The alternative is
controlled burns. But such fires would of course not be natural. Naturally occurring fires, on the other hand, need to be stopped when
there is a danger that they may spread to inhabited areas.[5] And to com-
plicate matters further, even those former fires of low intensity may
have been set by humans, the aboriginal inhabitants of this continent.

At the same time, noxious weeds like spotted knapweed and leafy
spurge that humans have imported from Eurasia are moving into the
wilderness. Unchecked they will suppress a variety of native plants
over large areas. Eradication and total suppression would be prohib-
itively expensive. To what extent and by what means should one control the spread of these weeds?

No wilderness area in the Lower 48 states is a self-regulating wildlife system. The number of elk depends on the availability of winter
forage, and the latter is often outside the wilderness boundaries or
specially managed for feeding. Thus the number and the movement
of the elk come to be determined by those areas that are set aside and
secured for them. The composition of elk herds as regards age and sex
is determined by hunting regulations. Other species such as the
fisher, the grizzly, and the wolf were once at home in the wilderness
areas. In most cases it will depend on our policies whether or not they
will once more be found there.

McKibben is pessimistic about a revival of nature. But the pessi-
mist, unlike the cynic, has some hope, however little.[6] The goal of
McKibben's "humbler world" is still the return of "an independent,
eternal, ever-sweet nature."[7] Nature in this sense is mortally
wounded because we "have changed the atmosphere, and thus we are
changing the weather. By changing the weather, we make every spot
on earth man-made and artificial."[8] The result that McKibben warns

of is this: "The world outdoors will mean much the same thing as the world indoors, the hill the same thing as the house."⁹ If independence is the mark of the truly natural, then, considering the fate of the wildest parts of this country, we must recognize that a restoration of the atmosphere would not revive nature. Must we surrender to the cynical view that everything is and ever will be artificial?

There is a more hopeful prospect. And it comes into view when we recognize how constricting the common distinction between the natural and artificial is. This distinction is seen entirely from the modern side of the postmodern divide that we are presently approaching. The restriction of the modern point of view is particularly clear from a closer consideration of the natural while the examination of the artificial will begin to open up the postmodern condition. To begin with, nature as McKibben and many environmentalists think of it in its healthy condition is characterized by its independence. It is unaffected by humans. This view sees the arena of reality just as the modern project sees it except that the environmentalists cheer the opponent of the modern attempt at domination. Whereas the proponents of the modern project used to reproach nature for its recalcitrance, the environmentalists had been hoping for its invincibility, and seeing their favorite threatened with defeat, they want to restore at least its independence.

Independence is perhaps the clearest criterion that has been used to define and save nature on the assumption that the world is basically controllable. There are other and perhaps more complex attempts to erect absolute criteria that would delimit and secure the natural environment once and for all. Among these norms are biodiversity, genetic variety, ecosystem, biocentrism, the intrinsic value of nature, and "the integrity, stability, and beauty of the biotic community."¹⁰

This view of the world is in the thrall of the modern project, not only in its conception of nature but also in its attitude toward the controllable and artificial complement of nature. In contemporary culture, we appreciate control because it rewards us with the pleasures of consumption. Yet the value of a life devoted to consumption is very much in doubt. Having defined nature as independent and having so removed it from people's daily commerce with their world, environmentalists have been unable to draw on their understanding of nature to clarify our doubts about consumption. Environmentalists do of course object to the recklessness of consumption and invoke the threat to nature in doing so. In this way nature becomes the source of an entirely cautionary and scolding sort of attitude. Consumption remains the sweetly attractive if reckless center of life.

Granted, McKibben has in addition to his practical concern with human survival a loftier one. He would like to retain his reverence of nature's unreachable independence or even transcendence. There is for him a close affinity between the independence of nature and the transcendence of God.[11] But this childlike awe is being undermined by the apparent end of nature and death of God. When McKibben is faced with the prospect of having to give up this kind of respect, he is deeply dismayed.

While once we humans were as children over against nature in its exceeding force, we are now as young adults, entrusted with the care of parents whom we have surpassed in physical and perhaps in mental power as well. Once they took care of us; now we have to be their care-takers. Accordingly, in the words of Walter Truett Anderson, we have to become "caretakers of a planet, custodians of all its life forms and shapers of its (and our own) future."[12] McKibben responds with a cri de coeur: "This intended rallying cry depresses me more deeply than I can say. That is our destiny? To be 'caretakers' of a managed world, 'custodians' of all life?"[13] I am sure many adolescents have cried out similarly when they saw their parents change from haven and refuge to task and burden, and some, terrified like Peter Pan, resolutely re-fuse to leave the charmed world of childhood. An outright refusal to outgrow the modern period can be seen in the rage that has been aimed at "The Big Lie" of the human restoration of nature.[14] I sug-gest, however, that we must overcome Peter Panic, accept Peter's commission—the keys to the kingdom—and, in the spirit of Eric Higgs's cheerful attitude toward restoration, set out to cross the post-modern divide.[15]

From the despairing modern point of view, the world entire ap-pears to be artificial and controllable. But a more forward-looking approach reveals a rather more complex and contingent world. To see this, imagine yourself immersed in the wintry reality of the Northern Rockies. A powerful way of experiencing it is alpine skiing. In the higher reaches of a ski area, you find yourself in a beautiful and forbidding world, and at seventy-five hundred feet you are the only charismatic megafauna in sight. The bears are hibernating; the cats and ungulates have descended to five or four thousand feet. The trees have been transformed into snow sculptures. You may come across a weasel, scurrying in and out of the snow, or a snowshoe hare flitting from one bush to another. Otherwise the chickadees in the trees and the crows in the air have the high country to themselves, a world extending endlessly, austere in whites and blues and, but for the peaks and ridges, soft and smooth of shape. Skiing down, you dive, bank, swoop, and turn much like the crows overhead. The world's center of gravity has shifted to these high and pristine slopes, and you are the animal that has the skill and grace to appropriate them fully.

But wait. How did you get up here? And what are you carving your turns on as you cruise down the hill? A high-speed chair lift scooped you up, rushed you along, and deposited you gently. Now you are flying down a run that has been cleared of trees and rocks, reshaped by bulldozers, and planted in grass. Underground there are miles of lines for water and compressed air, connected to snowguns that line the side of the run. At the bottom of the hill, a pumphouse and a compressor building supply water and air that, guided and monitored by computers, are mixed by the guns into the quality and quantity of snow needed at the time. It has taken a $20 million system with a thousand snow guns to produce the snow at a cost of $2,700 per acre-foot. But this is not all. An army of snowcats, $150,000 apiece, has worked all night to groom the slope to the shape of an undulating corduroy-surfaced ballroom floor.[16]

Natural snow has become dispensable—or, rather worse at times, a nuisance, as a Vermont marketing director tells us: "It sounds silly, but I hope it doesn't snow tomorrow. It will just make it difficult for people to drive up. The skiing we've got is already wonderful."[17] The cost of providing artificial snow, the crucial role that computers play in this ("many of the technological leaps can be traced directly to computers," says Steve Cohen), and the fact that people may still have problems getting to the winter wonder world all suggest that we should simply and bravely face up to what McKibben has told us is already the case: nature outdoors in essence is no longer distinguishable from the artificial indoors.[18]

Let me therefore make this modest proposal: an artificial indoor ski area in downtown Los Angeles. What would it look like? You may have seen in a sporting goods store the moving carpets that are mounted like large tilted conveyor belts and allow a skier to ski down the incline so that the skis sliding down and the carpet moving up roughly balance and, to a stationary observer, the skier stays in place. In addition to boots, skis, and poles, the skier is given a pair of goggles (skiers are used to these) where the lens is replaced by two microtelevision screens.[19] The rest of the story tells itself. We play on those screens moving scenes of ski slopes that are coordinated with the varying speed and pitch of the conveyor belt carpet. Everything else is a matter of technological refinement: blowers to simulate the rushing of the wind, a harness to suspend the wayward or crashing skier, and more. And let me briefly extol the virtues of the new kind of skiing, the reduction of gasoline consumption and automobile pollution, the infinite variety of conditions and terrains, the instant, continuous, and wide availability of skiing, and the supreme safety of the sport.

While the modern view of this pervasively artificial situation tends to inspire melancholy, as it has done in McKibben's experience, there is a postmodern reaction that welcomes the disappearance of the dichotomy between the natural and the artificial and indeed all dichotomies. Such a postmodern view revels in the resulting abundance of possibilities and is zealous in declaring all of them to be equally valuable. But there is an alternative and more sober postmodern vision that is concerned to point up crucial distinctions in the contemporary situation. These differences are best described, however, not as degrees of artificiality but as degrees of reality— "reality" taken in the sense of genuineness, seriousness, or commanding presence, the sense we have in mind when we speak of real gold as opposed to things that merely glitter and of a real person, a mensch, as opposed to a dude.

The philosophical challenge, of course, is to circumscribe this sense of reality in a way clear and precise enough to counter the suspicion of deconstructive postmodernism that advocates of a substantial reality are wistful and sentimental at best and patriarchal and fascist at worst. To fix our attention on a particular instance, how do we explicate the difference between a mountain in the Northern Rockies, covered with natural snow, and a skiorama in Los Angeles?

The difference, I suggest, is this. The mountain possesses a commanding presence and a telling continuity with the surrounding world. The skiorama, to the contrary, provides a disposable experience that is discontinuous with its environment. Consider first the experience of the mountain in winter. The snowy trail you are skiing down tells you about the particular and unsurpassable world you are in. This is January at seventy-five hundred feet. The amount of snow tells you something about the immediate past: Pacific moisture and arctic cold have been colliding and mixing with each other often this winter, but not often enough to produce a snowpack and a runoff this spring that would relieve the drought of these past six years.

The downhill skiing experience in the skiorama, being entirely at your disposal and discontinuous with its environment, tells you nothing about the world at large. It may in fact positively mislead you if you surrender to the sensation of cruising down a snowy slope on a bright winter morning. When the rolling carpet slows down and levels off and you remove your goggles, you are rudely returned to the sweltering midnight of Los Angeles.

Strangely enough, such artificial experiences, while doubtful in retrospect, seem superior to real ones in prospect and while in progress. They are more assuredly available and in greater diversity and, if engineered with care and sophistication, tickle our sensibilities more gratifyingly than real things ever could. This superiority is well captured in the term "hyperreality." Accordingly we can now say that today the critical and crucial distinction for nature and humans is not between the natural and artificial but between the real and hyperreal.

Though "hyperreality" is not a widely used term, the norm it is intended to capture is a powerful cultural force. It appears to have caught up with the fictional skiorama in Los Angeles. A recent AP story tells us of an "Indoor Mountain: Japanese Build Enclosed Ski Slope Near Tokyo," and it begins its account with the unmistakable intonation of hyperreality: "It sounds like the perfect ski resort: virgin snow every day, no wind or rain, easy to get to and no long lines for the lift."[20] When it continues with an enumeration of the shortcomings of the indoor ski arena—the lack of mountain scenery and fresh air—it merely points out what obstacles remain on the way to full hyperreality.

"Reality" in the widest of its senses refers to all there is, and to call something real in that sense is to say nothing. But we do in ordinary discourse use "real" in an eminent sense to pick out things and events that are notably serious, genuine, and valuable. One way of explicating the intuition that guides us in such talk is to say that what is eminently real has a commanding presence and a telling and strong continuity with its world. These two traits are connected. Whatever is devoid of contextual bonds and hence freely, that is, instantaneously and ubiquitously, available is therefore subject to our whims and control and cannot command our respect in its own right. Conversely, whatever engages our attention due to its own dignity does so in important part as an embodiment and disclosure of the world it has emerged from.

The distinction between (eminent) reality and hyperreality is not structured and secured by a bright dividing line that allows one to place whatever one comes across unfailingly on one side of it or the other. The distinction is moored by clear cases at the endpoints of a spectrum: wilderness at one end, for example, and videos at the other. In between there are intricate and interesting intermediates. A distinction is helpful if it provides orientation—and a continuum, firmly anchored at its extremes, does this as well as a dichotomy.

Still, for some social critics the real/hyperreal distinction is one without a difference or, at any rate, without a moral difference. By the most widely discussed moral standards, hyperreality does no worse and may very well do better than reality. In our example, the hyperreal instance appears to be more socially just, culturally diverse, environmentally benign, and physically healthy. But people of good sense have noticed that as we allow ourselves to slide from the real to the less real and hyperreal, our moral condition is undergoing a crucial change.

In the case of downhill skiing, this change has been well articulated by Lito Tejada-Flores. Contemplating the question whether skiing is better or worse than it was twenty years ago, he concludes:

> Skiing has not merely changed randomly under the influence of technology and marketing, it has evolved steadily in one direction. Evolved in three steps from adventure, to sport, to recreation. Maybe I'm being unfair to say this, but I see this evolution as the progressive reduction (and occasional elimination) of levels of challenge in skiing. Skiers have gone from *adventure* (dealing with uncertainty in a wild mountain landscape) to *sport* (all-out physical involvement on known terrain) to *recreation* (the undemanding enjoyment of simple rhythmic movements).[21]

Lito Tejada-Flores has seen that there is a symmetry between reality and humanity—that human nobility declines when the uncertainty of the wild mountain landscape is tamed and its challenge is silenced.

Lito Tejada-Flores has captured the heroic case of mountains and adventures. A child named Lucy has seen that the same decline is to be found in the inconspicuous details of contemporary culture. In a letter to God she writes: "Dear God, do plastic flowers make you mad? I would be if I made the real ones. Lucy."[22] Lucy sees what many of us have become blind to: that there is human impertinence in the refusal to care for what is fragile and to delight in what is passing—real flowers. Thus when the natural/artificial distinction is replaced by the real/hyperreal distinction, it is clear that the problem with consumption is not its sweet recklessness but its debilitating mindlessness.

But we need to grasp more firmly what follows from the postmodern condition for reality, nature, and human conduct. To begin with reality, its character at the close of the modern era is characterized by contingency. Most prosaically the term signals the discovery that reality is far less controllable and predictable than we have thought. We have come to see that physical processes exhibit patterns that are intricate beyond the modern dreams of computability and predictability.[23]

In a cultural sense, the contingency of reality means that the world refuses to comply with a vision of order that is fixed by a priori universal standards, be they the independence of nature, the transcendence of God, or the autonomy of humankind. But in neither the physical nor the cultural sense does contingency mean anything like featureless randomness.[24] Contingent reality has its own physiognomy, and the numerous and intricate lines that shade and separate off the real from the hyperreal constitute the most characteristic features on the face of the postmodern world.

What bearing does the contingency of reality have on the reality of nature? The principal task today is not to single out nature by some exclusive definition, but to include and appreciate it among the real and eloquent things and practices that are threatened by the hypertrophic overlay of hyperreality. This, of course, does not mean that the distinctiveness of nature should be submerged and lost in the contingency of reality. Something like "plants and animals" is sufficient to point up the natural to a first approximation, and from there it is a matter of distinguishing degrees of reality within nature.

The kind of reality that is akin to nature at its most real consists of the things and practices of athletics, of the arts, and of religion. What real nature today has in common with these is its powerful presence and a vigorous continuity with the world at large. Like athletics, art, and religion, nature today speaks in manifold and unforethinkable voices and, most important perhaps, in voices that are always responses to our own. And just as we would not doubt the autonomy of a spouse whose speech recalls the words of his partner, so we should not think of nature as broken simply because everywhere it now shows the traces of human actions.

Nature in this country speaks not only in the ancient voice of the wilderness but in its domestic intonations as well.[25] Vicki Hearne has lately reminded us that, although obviously we breed and train dogs and horses, they in turn can instruct and even humble us with their insight, courage, or elegance.[26] Nature can be eloquent in parks and gardens and can speak at the very center of our houses and apartments when it is celebrated in the culture of the table.[27]

What finally follows from the contingency of reality and the eloquence of nature for human conduct? Here too the reply to nature must be seen in the context of the human response to those real things that command our best efforts and orient us within the world at large. What they inspire us to do is to pursue the kind of excellence that culminates in celebration and is warranted by it. Such excellence is not the privilege of the rich or smart. There were times in human history when as a matter of common practice parents would teach their children the skills that issue in festive dining and dancing, in music and in storytelling. The nobility of practices like these is within everyone's reach.

The kinship of art and nature can further serve us as a guide to a clearer understanding of the standards of excellence we should aim at in our dealing with nature. Nature is now entrusted to us much like medieval cathedrals are to Europeans. Cathedrals are constantly abraded by the wear of time and the offenses of technology. To keep the destructive fumes and liquids of industry at bay takes determination. But that is not enough. Are we to let the spires, pinnacles, statues, and crockets erode into featureless remnants of their former glory? And if we remove the originals to safety and replace them with copies, are we content to see the cathedral turn more and more into a modern duplicate of itself? Are we willing to replace old timbers with steel girders, bell ringers with electric motors, resonant voices with public address systems?

There cannot be universal rules or algorithms for such problems. To each question an answer must be drawn from experience for this particular situation. And yet from the contingency of circumstances and efforts, the cathedral continues to rise in its own right. In this country, where no medieval bequests have come down to us, we have, belatedly and undaunted by the scorn of art historians, erected our own gothic churches. By now they have taken root in our communities and memories and become sacred sites.

The rightness of all this struggling and temporizing is warranted when people gather in cathedrals to celebrate. They may do so in the penumbra of divinity when, as secular citizens, they come to hear an organ concert. Or they do so as true believers when they attend mass. Music makes piers soar and vaults arch. Naves, transepts, and choirs center worship.

Similarly, every unroaded area needs to be secured, every wilderness attended to, and many an abused stretch of prairie or river restored or built up. Each such task requires its own approach and solution. If there is a general guideline, it would only be this: to save or restore the area's commanding presence and to guard its coherence with its environment and its tradition. And all that labor is warranted when we hike, ski, or canoe through a wild or natural area, real persons in real nature.

Human conduct that is invigorated by reality and devoted to excellence and celebration differs notably from conduct that is dedicated to the production and consumption of hyperreal commodities. Even if the latter pursuit were to live up to the conventional moral norms of ecological prudence and social justice, it would constitute a pyrrhic moral victory, for the result would merely guarantee that the debilitating mindlessness of consumption would be secured for all times and shared equally with all people.

But of course we in the United States are not as prudent as we might be nor nearly as just as we ought to be. Our selfishness here and among nations needs little elaboration. Thus it seems reasonable to suggest that even the single-minded proponents of ecological prudence or social justice have reason to pause and wonder whether their causes may not need the invigoration of a more substantial notion of moral excellence. Among the eloquent things that may invigorate and inspire us in this way, nature has a special standing in this country. We are uniquely Nature's Nation.[28] Nature in its many voices speaks more powerfully here than in Europe or Japan. If the people of this country learn to listen to those voices more attentively, they may regain that relaxed energy, generosity, and optimism—the grace under pressure—that other nations used to admire in this one.

Notes

1. See, for example, James Welch, *Fools Crow* (New York: Viking, 1986).

2. Carolyn Merchant, *The Death of Nature* (New York: Harper, 1983).

3. Bill McKibben, *The End of Nature* (New York: Random House, 1989), pp. 3–46.

4. John C. Hendee, George H. Stankey, and Robert C. Lucas, *Wilderness Management*, Miscellaneous Publication 1365 (U.S. Department of Agriculture, Forest Service, 1978), p. 82.

5. Hendee et al., *Wilderness Management*, pp. 249–278.

6. McKibben, *The End of Nature*, p. 215.

7. Ibid., pp. 190 and 209.

8. Ibid., p. 58.

9. Ibid., p. 48.

10. Aldo Leopold, *A Sand County Almanac* (London: Oxford University Press, 1949), pp. 224–225. For the intrinsic value of nature, see the *Monist*'s special issue on this topic, vol. 75, no. 2 (April 1992), with contributions by John O'Neill, Robert Elliot, Tom Regan, Eugene C. Hargrove, Bryan G. Norton, Jim Cheney, Anthony Weston, and Holmes Rolston III.

11. McKibben, *The End of Nature*, pp. 71–84.

12. Quoted in McKibben, *The End of Nature*, p. 214.

13. Ibid.

14. Eric Katz, "The Big Lie: Human Restoration of Nature," *Research in Philosophy and Technology* 12 (1992):231–241.

15. Eric S. Higgs, "A Quantity of Engaging Work to Be Done: Ecological Restoration and Morality in a Technological Culture," *Restoration and Management Notes* 9 (Winter 1991):97–104. For Peter's commission, see Matthew 16:18–19.

16. Steve Cohen, "High-Tech Snow," *Ski* (March 1992):28–39.

17. Quoted in Cohen, "High-Tech Snow," p. 28.

18. McKibben, p. 29.

19. Gary Stix, "Headsets: Television Goggles Are the Vision of the Future," *Scientific American* (March 1993):141.

20. Beth Sutel, "Indoor Mountain," *Missoulian* (22 June 1993):A8.

21. Lito Tejada-Flores, "Is Skiing Better or Worse Than It Was Twenty Years Ago?" *Powder* (January 1992):57.

22. An apocryphal letter in the tradition of *Children's Letters to God*, ed. Eric Marshall and Stuart Hample (New York: Pocket Books, 1966).

23. James Gleick, *Chaos: Making a New Science* (New York: Viking, 1987).

24. N. Katherine Hayles, *Chaos Bound: Orderly Disorder in Contemporary Literature and Science* (Ithaca: Cornell University Press, 1990).

25. Gary Snyder, *The Practice of the Wild* (San Francisco: North Point, 1990), and Wendell Berry, *What Are People For?* (San Francisco: North Point, 1990).

26. Vicki Hearne, *Adam's Task: Calling Animals by Name* (New York: Knopf, 1986).

27. Robert Farrar Capon, *The Supper of the Lamb* (Garden City: Doubleday, 1969).

28. That this is not a simple or easy fate is clear from Perry Miller's *Nature's Nation* (Cambridge: Harvard University Press, 1967).

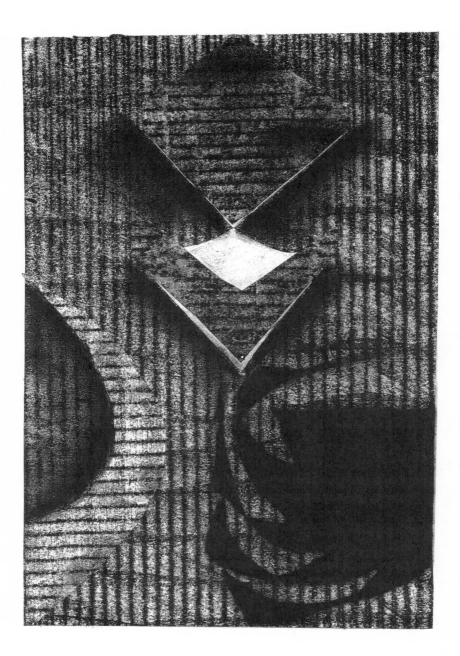

CHAPTER FOUR

SEARCHING FOR COMMON GROUND

N. KATHERINE HAYLES

A colleague recently told me of a talk he gave to a group of environmentalists about his work on Newton and the rise of seventeenth-century science. My friend espouses radical constructivism. He believes that everything we think we know, including "nature," is a construction emerging from historically specific discursive, social, and cultural conditions. The environmentalists were upset at this claim. The meeting took place at a small research station on a fog-shrouded island in Puget Sound, and they were incredulous that anyone could be amidst such breathtaking natural surroundings and still believe unmediated nature was not within our ken. Moreover, they worried about the effect the claim would have on environmental movements. If nature is only a social and discursive construction, why fight hard to preserve it?

Scientists find the claim upsetting for different reasons. They believe it threatens the very foundations of science, for it seems to imply that science does not play a privileged role in discovering the truth about reality. If scientific theories are merely social constructions, is science not trapped in a self-reflexive circle, mirroring the assumptions of its day and unable to reach beyond them? In this case, why undertake the demanding work of scientific investigation at all? Why not write novels or meditations instead?

My problem in sorting out these issues is that I identify with all three positions. Like the environmentalists, I have experienced the feeling of communing with nature; as a scientist, I have worked in a laboratory and been convinced that some models are more viable than others; and along with other constructivists, I have written and taught on the correlation between cultural conditions and changing ideas of what counts as science, nature, and reality. Searching for common ground for these diverse positions is for me not just a matter of bringing into conversation different disciplinary orientations (although that is part of what is at stake); it is also a question of working out a basis for action and thought that will satisfy the different voices within me. From amidst this heteroglossia, I think it is possible to glimpse an answer—or at least the shape of an answer—that responds to the legitimate concerns of the scientist and environmentalist and still holds onto the important insights articulated by social constructivism.

This answer is emerging from a broad spectrum of contemporary theory, including feminist challenges to scientific objectivity, the rethinking of the importance of embodiment among certain cognitive scientists, the emphasis on interconnectedness in ecology, and the recognition in anthropology of the complex ways in which physical environments, embodiment, discourse, and ideology collaborate to create a world. It finds concrete expression in recent advances in modeling nonlinear dynamic systems. Diverse as these approaches are, they share a common emphasis on interaction and positionality. Interactivity points toward our connection with the world: everything we know about the world we know because we interact with it. Positionality refers to our location as humans living in certain times, cultures, and historical traditions: we interact with the world not from a disembodied, generalized framework but from positions marked by the particularities of our circumstances as embodied human creatures. Together, interactivity and positionality pose a strong challenge to traditional objectivity, which for our purposes can be defined as the belief that we know reality because we are separated from it. What happens if we begin from the opposite premise, that we know the world because we are connected with it?

From Objectivism to Interactivity

Deeply rooted in Western culture, objectivism continues to form the backbone of the mainstream view of scientific inquiry. It has dominated the thinking of investigators as diverse as Descartes and Norbert Wiener, Francis Bacon, and Roger Penrose. Mark Johnson sums up its central assumptions: "The world consists of objects that have properties and stand in various relationships independent of human understanding. The world is as it is, no matter what any person happens to believe about it, and there is one correct 'God's-Eye-View' about what the world really is like."[1] Existing apart from humans, the world is grasped through correct reasoning, a cognitive property of the conscious mind. The mind, moreover, is conceived as a reasoning machine unaffected by the chassis in which it is mounted. The positionality of the human mindbody is thus largely erased, as are the language, culture, and belief systems of the observer.

The approach I am advocating maintains that the particular form of our embodiment matters, for it determines the nature of our interactions with the world.[2] Far from interfering with scientific inquiry, interaction is the necessary precondition for acquiring any knowledge about the world at all. But "world" is not the right word here, since it refers to a reality already constructed by our specific kind of sensory equipment, neural processing system, previous experience, present context, and horizon of expectation. Suppose that we think about what is "out there" as an unmediated flux. The term emphasizes that the flux does not exist in any of the usual conceptual terms we might construct (reality, nature, the universe, the world) until it is processed by an observer. It interacts with and comes into consciousness through self-organizing, transformative processes that include sensory, contextual, and cognitive components. These processes I will call the cusp.

For humans, the cusp is constituted through modalities peculiar to our physiology, including binocular vision, vertical posture, bilateral symmetry, apprehension of that portion of the electromagnetic spectrum we call light, and so forth. The cusp is also constituted through individual history and cultural expectations. Research by Christine Skarda and Walter Freeman, for example, has shown that for many species, including humans, previous experience influences how sensory data are processed, even before they are consciously apprehended.[3] To illustrate our complete dependence on the cusp for our understanding of the world, I have speculated elsewhere on how Newton's three laws might look to a frog.[4] Drawing on the famous article by J. Y. Lettvin and colleagues, "What the Frog's Eye Tells the Frog's Brain,"[5] I argued that a frog gifted with Newton's reasoning power but with a consciousness constituted through a frog's sensory equipment would have drawn very different conclusions than Newton did from being hit on the head with an apple. Of course, such an idea is a playful oxymoron, because my argument implies that cognitive abilities are intimately bound up with the sensorium that helps to constitute them. To be a frog is to think like a frog, no less than to be a human is to think like a human. As Humberto Maturana (one of the investigators who did the pioneering work on the frog sensorium) was fond of pointing out, there can be no observation without an observer, including this observation itself.[6] The reflexivity inherent in this situation suggests that Newton's laws (and any other scientific generalizations) are not true in any absolute or transcendent sense removed from a human context. Rather, they are consistent with the experiences of species-specific, culturally formed, and historically positioned actors.

On one side of the cusp is the flux, inherently unknowable and unreachable by any sentient being. On the other side are the constructed concepts that for us comprise the world. Thinking only about the outside of the cusp leads to the impression that we can access reality directly and formulate its workings through abstract laws that are universally true. Thinking only about the inside leads to solipsism and radical subjectivism. The hardest thing in the world is to ride the cusp—to keep in the foreground of consciousness both the active transformations through which we experience the world and the flux that interacts with and helps to shape those transformations. There is a constant temptation to forget the complexities of the cusp and abstract to the shorthand of a world that exists independent of our interaction with it. From such abstraction comes the belief that nature operates according to laws that are universally and impartially true.

What is the harm in moving to the abstraction? The implications become clear when we look at what it leaves out of account. Gone from view are the species-specific position and processing of the observer; the context that conditions observation, even before conscious thought forms; and the dynamic, interactive nature of the encounter. In such a pared-down account, it is easy to believe that reality is directly accessible, that interaction is nothing more than an additive combination of individual factors, each of which can be articulated and analyzed separately from the others, and that the observer can be extracted from the picture without fundamentally altering the picture itself.

This is, of course, the world of classical physics. It continues to have a vigorous existence in popular culture as well as in the presuppositions of many practicing scientists. When the TV camera, accompanied by Carl Sagan's voice-over, zooms through the galaxy to explore the latest advances in cosmology, these presuppositions are visually and verbally encoded into an implied viewpoint that seems to be unfettered by limitations of context and free from any particular mode of sensory processing. As a representation, this simulacrum figures representation itself as an inert mirroring of a timeless, objective reality.

Perhaps the most pernicious aspect of the objectivist view is the implicit denial of itself as a representation. The denial is all the more troubling because of the ideological implications encoded within it. Among those who have explored these implications are Evelyn Fox Keller, who points out the relation between an "objective" attitude, the masculine orientation of science, and the construction of the world as an object for domination and control;[7] Ilya Prigogine and Isabelle Stengers, who relate the appeal of a timeless realm to a fear of emotional involvement and death;[8] and Sandra Harding, who underscores the differences between non-Western and Western science to make visible the assumptions within the dominant traditions of Western scientific practice.[9] These critiques can be seen as acts of recovery, attempts to excavate from an abstracted shorthand the complexities that unite subject and object in a dynamic, interactive, ongoing process of perception and social construction.

A model of representation that declines the leap to objectivity figures itself as species-specific, culturally determined, and dependent on context. Emphasizing interactions rather than disembodied observations, it insists that embodied experience constructs a world, not the world. Recognizing that speakers and actors are always positioned (through embodiment, culture, language, and a host of other historical specificities), it indexes its conclusions to the contexts in-which implied judgments about accuracy are made. Yet it also acknowledges that not every claim is equally valid. Unlike many versions of cultural relativism, this position maintains we can determine whether some statements about our interactions with the flux are more valid than others.

Since the insistence that not all claims are equally valid separates this position from strict social construction as well as from cultural relativism, it is worth exploring more fully. Central to this position is the idea of constraints. By ruling out possibilities, constraints enable scientific inquiry to tell us something about reality and not only about ourselves. Consider how conceptions of gravity have changed over the last three hundred years. In the Newtonian paradigm, gravity is conceived very differently than in the general theory of relativity. For Newton, gravity resulted from the mutual attraction between masses; for Einstein, from the curvature of space. One might imagine still other kinds of explanations—for example, a Native American belief that objects fall to earth because the spirit of Mother Earth calls out to kindred spirits in other bodies. No matter how gravity is conceived, no viable model could predict that when someone steps off a cliff on earth, she will remain spontaneously suspended in midair. Although the constraints that lead to this result are interpreted differently in different paradigms, they operate universally to eliminate certain configurations from the realm of possible answers. Gravity, like any other concept, is always and inevitably a representation. Yet within the representations we construct, some are ruled out by constraints and others are not.

The power of constraints to enable these distinctions depends on a certain invariability in their operation. For example, the present limit on silicon technology is a function of how fast electrons move through a semiconductor. One could argue that "electron" is a social construction, as are "semiconductor" and "silicon." Nevertheless, there is an unavoidable limit expressed in this formulation about an electron's speed, and it will manifest itself in whatever representation is used, provided it is relevant to the representational construct. Suppose that the first atomic theories had developed using the concept of waves rather than particles. Then we would probably talk not about electrons and semiconductors but indices of resistance and patterns of refraction. There would still be a limit, however, on how fast messages could be conveyed using silicon materials. If both sets of representations were available, one could demonstrate that the limit expressed through one representation (electrons as particles) is isomorphic with the limit expressed in another (electrons as waves).

I am not saying constraints tell us what reality is. They cannot say "This is true," for truth implies that we can occupy some position from which we can see reality-in-itself and can therefore verify that our representation matches it. The most constraints enable us to say is "This is consistent with our interactions with the flux." Three aspects of this statement are worth emphasizing. First, the representation is matched with our interactions with the flux rather than with reality-in-itself, so that positionality is built into the formulation from the ground up. Second, the claim of truth is replaced by the claim of consistency. We do not need to see reality-in-itself to know whether a representation is consistent with our experience; we need only to know the experience. Third, the statement is capable of negation; that is, it is also possible to say: "This is not consistent with our interactions with the flux." Thus some claims can be eliminated as inconsistent. By enabling this distinction between models consistent and inconsistent with our experience, constraints play an extremely significant role in scientific research, especially when the representations presented for disconfirmation are constrained so strongly that only one is possible. The art of scientific experimentation consists largely of arranging situations so the relevant constraints operate in this fashion. No doubt there are other representations, unknown and perhaps for us unimaginable, that are also consistent with reality. The representations we present for falsification are limited by what we can imagine—which is to say, by the prevailing modes of representation within our culture, history, and species. But within this range, constraints can operate to select some as consistent with reality, others as not. We cannot see reality in its positivity. We can only feel it through isomorphic constraints operating upon competing local representations.

The term I propose for the position I have been urging is constrained constructivism. The positive identities of our concepts derive from representation, which gives them form and content. Constraints delineate ranges of possibility within which representations are viable. Constrained constructivism points to the interplay between representation and constraints. Neither cut free from reality nor existing independent of human perception, the world as constrained constructivism sees it is the result of complex and active engagements between the unmediated flux and human beings. Constrained constructivism invites—indeed, cries out for—cultural readings of science, since the representations presented for disconfirmation have everything to do with prevailing cultural and disciplinary assumptions. But not all representations will be viable. It is possible to distinguish between them on the basis of what is actually there.

"Actually there?" Yet I have also argued that as situated and embodied creatures, we can know the flux only as it comes into existence through our interaction with it. In what sense do constraints function as conduits to the flux? Are they not themselves representations and therefore subject to the same limits as all other representations? It is possible, of course, to create a representation that points toward a constraint. The Second Law of Thermodynamics is such a representation. It states that in a closed system, entropy always tends to increase. Models that do not fulfill this requirement are considered to be thermodynamically impossible. The Second Law differs from the First and Third Laws because it has never been possible to derive it from first principles. Historically there has been some uneasiness about its status, since it seems more like an empirical generalization than a "law." In this respect it is like a constraint; it is also like a constraint in that it limits the range of viable models. Nevertheless, despite its similarity to a constraint, the Second Law is not a constraint in the sense that I have been using the term, for constraints by definition have no explicit formulation, no articulation as things-in-themselves. Not only can they not speak the positive truth about the world; they cannot speak themselves. A constraint that is expressed is a representation, not a constraint.

How then do we know what constraints are, or even if they are? We can infer that constraints exist because of the consistencies that obtain in our interactions with the flux. Like the Second Law, this belief remains an inference and a generalization rather than a "law" as such. Representations that point toward constraints can be formulated when we have multiple sets of representations to compare with each other; no closed system has ever been observed to result in a net decrease of entropy, just as no rock has ever spontaneously spun off the earth into space. Nevertheless, there always remains a necessary gap between hypothesizing the existence of a constraint and saying what it is. Constraints thus embody a double negative. They negate some representations as inconsistent with our interactions with the flux, and they negate the possibility that they can ever be fully articulated themselves. "Elusive negativity," Shoshana Felman has called this doubly negated semiotic position.[10] Precisely because elusive negativity has no definite shape of its own, Felman identifies it as the source of fertile new insights. In its very elusiveness it leaves the door open to unforeseen possibilities. If constrained constructivism means that we must abandon the illusion of a God's-Eye-View, it promises in exchange encounters with a world that is much more capacious than our views alone can imagine.

The Value of Interactivity

"Values" are too often treated in scientific discourse as if they were written not in the book of nature but in an appendix to it, added on afterward rather than intrinsic to the stories through which we constitute nature for ourselves and others. In fact, value-laden presuppositions operate from the moment one begins to speak or write. Take the instance I discussed earlier of different conceptions of gravity. Why gravity? The choice is not coincidental: in the numerous conversations I have had with scientists about social constructivism, gravity is invariably brought forward as the great counterexample showing that science is not culturally constructed. A rock falls to earth regardless of the dominant language or ruling class. Yet even the pervasiveness of this example indicates that it is culturally encoded, for it is linked to a specific history in which Newton's formulation of the law of gravity was an epoch-making event. More subtly, the example is marked by a certain cultural position because it presupposes that mathematics and physics are the core sciences rather than, say, biology and ecology. Many scientists (especially physicists and mathematicians) will say as much explicitly if pressed. They regard physics as the archetypal physical science because, as Norbert Wiener put it, physics manages largely to escape the "messiness" of "tight couplings" between the observer and the observed. The same presuppositions that inform objectivism as a philosophical position also create a hierarchy of sciences that places physics and mathematics at the top, home economics and animal husbandry near the bottom.

Shifting the focus to our connection with the world rather than our separation from it thus involves more than initiating a different epistemology. Inevitably, it constitutes a shift of values as well. Carolyn Merchant, among others, has written on the attitudes of domination that were bound up with the rise of science in the seventeenth century.[11] By separating subject from object, objectivism helped to constitute the belief that one could act upon the world without oneself being acted upon. The Baconian vision of human domination of the planet through science could thus be put forward as an enterprise that would not necessarily also affect those who dominated. Extrapolated to the environment, this attitude implies that rainforests can be cut without affecting those doing the cutting; rivers polluted without poisoning those polluting; fluorohydrocarbons released without affecting those doing the releasing. By contrast, a science understood to flow from historically specific interactions implies that we know the world because we are involved with it and because it impacts upon us. While such an understanding of the scientific enterprise does not guarantee respect for the environment, it provides a conceptual framework that fosters perceptions of interactivity rather than alienation.

Similarly, interactivity foregrounds rather than obscures the importance of embodiment. In the interaction model, the body does more than provide a biological support system for the mind. Interaction is possible only because we are embodied, and the precise conditions of our embodiment have everything to do with the nature of those interactions. The range and nature of sensory stimuli available to us, the contexts that affect how these stimuli achieve meaning, the habituated movements and postures that we learn through culture and that are encoded for gender, ethnicity, and class—all affect how learning takes place and consequently how the world comes into being for us. To be incorporated within a different body would be to live in a different world.

Objectivism, positing the body as a secondary phenomenon derived by the cogitating mind, fosters the illusion that we can somehow transcend our bodies and live in the realm of pure thought. In contemporary cognitive science, this belief achieves concrete expression in Hans Moravec's research program to download human consciousness into a computer.[12] In Moravec's view the body is mere wetware, a sticky and unhandy—not to mention mortal—medium that can be discarded once the transfer of mind into computer is complete. Scratch the surface of Moravec's rhetoric and that of likeminded researchers, and you find a deep anxiety about whether we can continue to live on this planet. When time-release environmental poisons make our continued physical existence problematic, it is not surprising that dreams of disembodied transcendence flourish. By insisting on the centrality of embodiment to human experience, interactivity discourages these illusions and vivifies the stakes we have as embodied creatures in contesting for the integrity of the environment.

Another arena in which interactivity has important ethical implications emerges from its emphasis on the species-specific nature of our representations. To imagine a world apart from any specific observer, objectivism has to posit a fundamental separation between world and subject. Alienation, far from being a by-product of objectivism, is interwoven into its constitutive presuppositions. "Nature" is an object of knowledge studied by subjects who gain knowledge. The knowledge belongs to the subjects, not to the objects. Animals, belonging to the world of "nature," are understood as objects separate and apart from the subjects who study them. A dog or a rat is an object which can be manipulated to produce knowledge, rather than a subject who himself knows. In a science inscribed with this ideology, it seems to make sense to sacrifice animals for the sake of objective knowledge, for no fundamental connection is posited between what we learn and what they experience as suffering subjects. The sum total of knowledge is considered to be increased, not diminished, by their sacrifice.

Interactivity does not so much overcome alienation as constitute a different kind of world in which objectivity can be seen as a misperception. In the interactive model, if one wants to think about the totality of the world, one can no longer achieve it by what Donna Haraway calls the "god trick."[13] Any view at all, much less a total view, is literally unthinkable without perceiving subjects to bring it into being through interaction with the flux. Different subjects—a human, a dog, a rat—construct different worlds through their embodied interactions with it. Each construction is positioned and local, covering only a tiny fraction of the spectrum of possible embodied interactions. How then does one reach toward the world's totality? By imaginatively bringing together the different knowings that all the diverse parts of the world construct through their interactions with it. Situated within our human perspectives, we move toward the unknown when we strive to comprehend how other creatures experience the flux. Although we can never completely or even adequately know that portion of the whole which they contribute, our attempts to make contact with it enrich our perspectives by hinting at how our positionality differs from theirs. We know ourselves by reaching toward the other. To sacrifice animals or exterminate species in this model directly reduces the sum total of knowledge about the world, for it removes from the chorus of experience some of the voices articulating its richness and variety.

As Carolyn Merchant has observed, it is no accident that certain practices and theories grow out of a given framework for knowledge construction, while another framework gives rise to different ideas and theories.[14] Although the framework does not uniquely determine the theories, it creates a matrix for thought that defines ranges of possibility. Changing the framework is always consequential, both for scientific theories and practices and for the larger social issues connected to them. The change I have been urging is a rejection of objectivism and an embracing of interactivity. To explore more fully how this model of knowledge construction can reinvent nature, I turn now to consider its other major component: positionality.

Positionality and the Power of Limits

The foregoing implies that we can never know whether a representation is true, in the sense that it is congruent with reality, because we have no exterior place to stand from which we can see reality as such. Hence we can never compare representations to reality, only to one another. I have suggested that constraints can eliminate certain representations from the range of viable options, but many times several different representations may be consistent with our interactions with the flux in specified contexts. The choice between the Ptolemaic and Copernican systems was of this nature. Each was consistent with available observations, for additional epicycles could always be added that made the Ptolemaic representation as predictive as the Copernican.[15] In such cases additional criteria have to be introduced, and these are always encoded with cultural assumptions. The principle of parsimony, for example, which says that the explanation requiring the least number of ad hoc moves is preferable, comes out of a tradition that privileges reduction over expansion. The principle of parsimony is not a fact of nature but a judgment, and in some situations—for instance, the study of complex systems—a judgment that has had negative effects on understanding natural phenomena.

The contexts in which judgments about the adequacy of representations are made, then, obviously include more than the species-specific nature of our perceptual processing, which itself exists as a spectrum rather than a single point. Language, history, culture, disciplinary tradition, gender, class, and race, among other factors, can all be relevant to the context of judgment, depending on the situation at hand. As Sandra Harding has pointed out, the scientific method operates only upon competing representations; it says nothing about how to generate representations or how to expand the range of representations available for testing.[16] When assumptions are widely shared within a discipline—when, say, the researchers are all or almost all white First World males—the range of representations can be very narrow relative to those that may be consistent with human interactions with the flux in different cultural contexts. Considering representations that emerge from other positions can be immensely valuable, for through their differences they reveal assumptions that have remained hidden to those who occupy the dominant roles.

A similar line of thought has led Sandra Harding to assert that traditional objectivity is flawed because it is not objective *enough*.[17] Instead, Harding argues for what she calls strong objectivity, which involves recognizing that one is always positioned and that this position affects what one sees, sometimes in very literal ways. The more one can understand about one's positionality, the more objective one can be in this strong sense. Those who do not recognize their positionality nearly always occupy positions of privilege, for what privilege means in this context is precisely having the power to ignore competing representations made from other positions. People in privileged positions should make special efforts to attend to constructions coming from other locations, this argument implies, not only or even primarily for reasons of justice and equity, but for the sake of their own objectivity and knowledge. The endeavor to learn more about non-Western science is thus as important to First World peoples as it is to those who occupy the margins.

What does the proposed view of knowledge construction, with its emphasis on interactivity and positionality, have to offer the scientist and the environmentalist with their concerns about social constructivism? For the scientist, it offers the reassurance that we interact with the unmediated flux and that this interaction is governed by constraints that limit which representations will be viable. There remains, of course, a difference between calling what is out there an "unmediated flux" and calling it an "underlying reality." Whereas the traditional expression implies that reality exists in some timeless, foundational way, the language of mediation emphasizes that the flux can come into existence for us only through our interaction with it and hence is already constructed. The view I have been espousing would agree with the scientist that models may be more or less adequate. It differs from the traditional view in leaving open the question of how language, culture, and historical position determine the models proposed for testing and affect the design and interpretation of testing procedures. In addition, it urges that representations emerging from marginal positions have a special role to play in knowledge construction.

For the environmentalist, this view argues that a constructivist position need not lead to a laissez-faire attitude toward the environment or encourage the worst excesses of postmodern society. On the contrary, it points out that human interactions with the flux comprise only an infinitesimal fraction of possible modes of being in the world. It extends our awareness of anthropomorphism into a perspective that values other species' encounters with the flux precisely because they are different than ours. Understanding that we are positioned, we also understand that we are limited and finite creatures. The unique contribution that wildness can offer us then becomes apparent, for as Gary Snyder vividly reminds us, the human perspective is only one voice in a rich chorus of experience.[18]

Finally, I should like to consider what this view offers social constructivism. Since my analysis is obviously deeply indebted to constructivism, it may seem strange that social constructivists could find it as offensive as do scientists and environmentalists. Yet it is possible to see the position I have outlined as conceding too much to realism, for it argues that not all representations are equal and that reality-based constraints can eliminate certain representations as possible candidates. Anyone familiar with the literature of social constructivism, particularly the strong program, can scarcely avoid noticing in it a strong strain of reflexivity. All too often, this reflexivity takes the form of anxiously noticing that the writer too has a social class and cultural position, which observation is also positioned, and so on into an infinite regress that seems to lead only into solipsism. The important work that reflexivity can do is opened up once it is associated not with relativism but with strong objectivity, for then it is an empowering strategy that connects positionality with expanding the range of what can be thought and imagined. Reflexivity, understood as recognizing that one has a position and that every position enables as well as limits, can make the double move of turning outward to know more about the world because it also turns inward to look at how one's own assumptions are constructed. If constrained constructivism does nothing more than enhance the humility we feel when we realize that the world is a much bigger place than we as situated human beings can imagine, in my view it is worth the price of admission.

Notes

1. Mark Johnson, *The Body in the Mind: The Bodily Basis of Meaning, Imagination, and Reason* (Chicago: University of Chicago Press, 1987), p. x.

2. This portion of my argument appeared in somewhat different form in "Constrained Constructivism: Locating Scientific Inquiry in the Theater of Representation," *New Orleans Review* 18 (Spring 1991):76–85, reprinted in *Realism and Representation: Essays on the Problem of Realism in Relation to Science, Literature and Culture*, edited by George Levine (Madison: University of Wisconsin Press, 1993), pp. 27–43.

3. Christine A. Skarda, "Understanding Perception: Self-Organizing Neural Dynamics," *La Nuova Critica* 9–10 (1989):49–60. See also Walter Freeman and Christine Skarda, "Mind/Body Science: Neuroscience on Philosophy of Mind," *John Searle and His Critics*, edited by E. LePore and R. Van Gulick (London: Blackwell, 1988).

4. "Constrained Constructivism," p. 76.

5. J. Y. Lettvin, H. R. Maturana, W. S. McCulloch, and W. H. Pitts, "What the Frog's Eye Tells the Frog's Brain," *Proceedings of the Institute for Radio Engineers* 47 (1959):1940–1951.

6. Humberto Maturana and Francisco Varela, *Autopoiesis and Cognition: The Realization of the Living*, Boston Studies in the Philosophy of Science, vol. 42 (Boston: D. Reidel, 1980).

7. Evelyn Fox Keller, *Reflections on Gender and Science* (New Haven: Yale University Press, 1985).

8. Ilya Prigogine and Isabelle Stengers, *Order Out of Chaos: Man's New Dialogue with Nature* (New York: Bantam, 1984).

9. Sandra Harding, "After the Neutrality Ideal: Science, Politics, and 'Strong Objectivity,' " *Social Research* 59 (Fall 1992):567–587.

10. Shoshana Felman, *The Literary Speech Act: Don Juan with J. L. Austin, or Seduction in Two Languages*, translated by Catherine Porter (Ithaca: Cornell University Press, 1983), pp. 141–142.

11. Carolyn Merchant, *The Death of Nature: Women, Ecology, and the Scientific Revolution* (New York: Harper & Row, 1989).

12. Hans Moravec, *Mind Children: The Future of Robot and Human Intelligence* (Cambridge: Harvard University Press, 1988).

13. Donna Haraway, "Situated Knowledges: The Science Question in Feminism as a Site of Discourse on the Privilege of Partial Perspective," *Feminist Studies* 14 (1988):575–599.

14. Merchant, *Death of Nature*.

15. Thomas Kuhn discusses this case at length in *The Structure of Scientific Revolutions*, 2nd ed. (Chicago: University of Chicago Press, 1970).

16. Sandra Harding, "Having It Both Ways: 'Strong Objectivity' and Standpoint Epistemologies," in *Feminist Epistemologies*, edited by Linda Alcoff and Elizabeth Potter (New York: Routledge, 1993).

17. Harding, "After the Neutrality Ideal."

18. Gary Snyder, *The Practice of the Wild* (San Francisco: North Point, 1990).

CHAPTER FIVE

NATURE AND THE DISORDER OF HISTORY

DONALD WORSTER

When I glance out my window, I see a rural Kansas landscape in an apparent state of stability. Events come and go but the composite remains stable to the eye year after year. A flock of wild turkeys may occasionally pass by, a windstorm may whip the trees violently, a snowstorm may blanket the ground and then the sun melt the snow; but the sky does not abruptly change places with the earth one morning nor a line of trees make a sudden advance toward the house. Yet I know there is change going on out my window and that the landscape presents more than an endless cycle of the same events.

My house sits in the midst of old marginal agricultural lands reverting to forest: native bur oak, locust, and walnut mixed with introduced Osage orange and Scotch pine. Compared to the changes coming over the radio, that shift in the landscape seems soothingly slow. Nature moves in new directions, but its rhythms of change are quite different from those in politics or economics or the recording industry. Nature changes, I acknowledge, but applying the simple word "change" to the landscape does not help me make any discriminations among the many kinds and rates of change going on around me.

As an environmental historian, I am supposed to be looking for a story of change to tell. Changes in people's attitudes toward the natural world have been among the most dramatic stories my field has described—the change "from mountain gloom to mountain glory," for example, or from John Muir's campaign for national parks to Rachel Carson's campaign against chlorinated hydrocarbons, or from an American culture of manifest destiny to one of bioregionalism. Environmental history has also told the story of a changing biophysical environment, altered by the forces of nature and technology working together in a complicated dialectic. We realize more than ever that those shifting attitudes toward the natural world are to some degree the product of that complex dialectic: as material conditions change, what is called "nature" disappears and its place is taken by a new construct.

One effect of this history of change is to call into question any naïve, romantic assumptions about a static world of unspoiled nature. Indeed, for most educated people such assumptions disappeared long ago. Today one encounters a widespread nostalgia for a wilder, less managed world but not much naïveté about a truly static nature. Yet one does not find much real knowledge about the historical environment either, or much careful grounding of values in historical possibilities. Many environmentalists, for example, are too caught up in current political battles to think deeply about what it is they want to save and about what kind of changes are acceptable in the landscape and what are not. A useful role of the historian, then, is to help people clarify the issues. Nature changes, we point out, even though that change is not always apparent to the eye. What we want out of nature also changes. What we want and what we get are never the same, either for environmentalists or for developers. What we wanted in the past had consequences that no one expected: surprises, unpredictable outcomes, many disappointments, some of which were tragic. Historians delve into all these issues and help to create an intellectual context for their society that defies simplistic thinking or unrealistic expectations.

Yet I have to admit that historians can also become a dangerous influence in environmental policy debates. Their passion for studying change can become a fixation that distorts reality. Life is full of changes, historians insist, but that insight does not make change the only, or even the most important, metaphysical principle on earth. We have not yet discovered what that ruling principle is, or indeed if there is one, so we are to some extent free to choose what we will emphasize: we can choose to focus our attention on change, or we can choose to focus on stability, persistence, resilience, and continuity. Historians tend to have a vested personal and professional interest in doing the former. They become promoters of a certain view of the world and may even be intolerant toward any other, insisting that change is all there is. Historians may want to relativize everything.

As professions go, historians are fairly harmless and inconsequential. Compared to, say, nuclear physicists or economists or lawyers, they have little money to spend or little power to exert over the councils of state. Nobody pays much attention to their monographs on banking regulation during long-forgotten presidencies or monastic life in the Middle Ages. But on another, more indirect level historians have exercised a great deal of influence; they have, through their insistence on exploring the past, taught the world to think historically—to search for origins, to focus on development, to track change over time. Now, no field of inquiry is immune to that perspective, so that it has become perhaps the most basic perspective of our times, underpinning all public policy, all science, all moral relations. Children grow up assuming that their parents lived in radically different times, when attitudes toward sex or school or leisure were unlike those of today. Likewise, the physical environment of past ages, we learn, must have been quite different from that of today, an environment filled with other creatures and strange people. We may believe that those past environments were better than our own in the sense of being less polluted, less crowded, more green, yet we have also been schooled to see that there have been no true golden ages, no examples of ideal harmony to be found in the past; if people once trod the earth more lightly than we do today, they did so by practicing a brutal infanticide or by suffering warfare, hunger, and disease that restrained their populations. In our education the awareness of change has become a leading principle, which is to say that we have all become historians.

The danger in this development lies in the excessive relativism that historical thinking can produce. It is a danger facing every aspect of modern life, but I am especially interested in its implications for conservation. If nature is nothing but a bewildering panorama of changes, many of them induced by human beings, going back to ancient hunters setting fire to the bush, and if our attitudes toward nature are themselves demonstrably in a state of constant flux, so that yesterday we hated wolves and today we love them, then what should conservation mean? What should we derive from the study of history to inform current land-use decisions? Is environmental history at all useful to the management of land, or is it a mental disease confusing decisions and clouding judgment?[1]

Before trying to answer these questions, we ought to examine how our picture of nature has been thoroughly historicized during the last two centuries, especially during the last two decades. Science itself has become a branch of history. How this happened is the theme of a book by Stephen Toulmin and June Goodfield entitled *The Discovery of Time*. "The picture of the natural world we all take for granted today," they point out, "has one remarkable feature, which cannot be ignored in any study of the ancestry of science: it is a *historical* picture."[2] The new picture began to emerge during the period 1810 to 1830 as scientists began to realize how much time had transpired on the earth and how much had changed over that span of time. A static Newtonian world of fixed, hierarchical relations began to give way to another nature: evolving, contingent, revolutionary, conflicted, catastrophic at times, always in a state of flux. Geology was the first science to discover time; the first great geologists, James Hutton, William Playfair, and Charles Lyell, were all historians, finding the annals of former worlds written in deeply buried beds of chalk and old red sandstone.

It is no coincidence that the modern academic discipline of history had its roots in that same era when deep time first began to be discovered. Like the newly discovered fossils lying embedded in the dust, waiting to be exhumed and analyzed, great political empires of the past had to be dug up and explained. A generation that had just been through many profound revolutions on three continents, Europe and both the Americas, could not help wondering how long it would be before the next upheaval came along. Thomas Jefferson, an earnest student of ancient empires, called for making a revolution in every generation. The future promised to be unlike anything ever seen before, and the past became its mirror, full of strange, exotic ruins demanding explanation. Historians like Gibbon, Macauley,

Michelet, Ranke, Bancroft, and Parkman began to write long, eloquent meditations on the meaning of the past.

Nor is it a mere coincidence that the same century which created modern history, which became fascinated by a very long human chronology, which discovered in the fossil record the traces of countless extinct species, saw the appearance of the theory of evolution through natural selection. Charles Darwin turned biology into history—the history of flora and fauna jostling for space, branching out to new territory, overthrowing established regimes. According to Toulmin and Goodfield, Darwin's book *On the Origin of Species* "broke down the artificial barrier between Science—which had hitherto been concerned with the static Order of Nature—and History, which studied the development of humanity. So the two most powerful intellectual currents in the nineteenth century were united. Whether we consider geology, zoology, political philosophy or the study of ancient civilizations, the nineteenth century was in every case the Century of History—a period marked by the growth of a new, dynamic world-picture."[3]

Science and history had become one interconnected story of nature and humanity moving forward toward perfection. Heretofore science had dealt with a nature that seemed timeless, impervious to human impact, obedient to a perfectly rational system of laws. History, on the other hand, had taken as its subject the far messier sphere of human life, with all its palace intrigues, its wars and debaucheries, its emotional explosions, its guillotines and mobs. Now the two spheres were joined together by a shared sense of life struggling and evolving through time. It was Charles Darwin who, more than anyone else, announced the union with the publication of his second book, *The Descent of Man* (1871), a biological history of the species that everyone understood to have immense cultural significance.

With Darwin, as with other thinkers of the century, historians as well as scientists, change is never all there is in nature. Change *leads* somewhere; it has a discernible direction, conventionally called Progress. Darwin described evolution as a great blooming tree of life, suggesting that change is coherent and contained, like the growth of an organism, where the parts increase or even replace one another but the whole remains one entity. Nature, like human society, told a story of constant upheaval, but the observer could still find order and pattern in the story.

Out of that new historicized biology came the field of ecology, though it was not until the 1890s that ecology could be said to have achieved a rudimentary disciplinary status. How could the new science be anything but historical, born as it was at the end of the great century of historical and developmental consciousness? Its founders, including Ernest Haeckel, Eugenius Warming, and early in the next century Henry Cowles and Frederic Clements, were all intensely aware of the biological and geological past, of time's arrow flying unstoppable. Like Darwin, however, they believed that change is not at all disorderly or directionless. Change unmakes order but also makes it anew. Despite a thousand mishaps, nature has its regularities, its great coherences that persist over time, giving the landscape a definition of normality or goodness.

This act of balancing a sense of historical change with a sense of stability, of finding within the swirl of history a normative state, persisted in ecology well into the 1960s. Eugene Odum was its last great exemplar, in the several editions of his textbook, *The Fundamentals of Ecology*, which laid so much stress on natural order that it came close to dehistoricizing nature altogether. For Odum ecology was the study of the "structure and function of nature," a definition that almost left out of the picture Darwinian evolution and all its turmoils. The ecosystem was made the key concept in the field, and change became little more than "the development and evolution of the ecosystem," which is to say the process of ecological succession. The process was orderly, reasonably directional, and predictable, and it culminated in stability, or what Odum called "maturity." His famous tabular model of succession, published in *Science* in 1969, was a map of change that had only two categories: "developmental stages" and "mature stages." The history of nature was by and large reduced to a movement from one category to the other, after which change normally came to an end. Ecosystems matured but they did not die. They reached a condition of near immortality.[4]

The only other kind of history that got much attention in Odum's work appeared briefly in a discussion of the changing relations of plant life to the global atmosphere. Drawing from the work of Helen Tappan in the new subfield of paleoecology, Odum acknowledged that in archaic times the planet had gone through tumultuous revolutions, including the invention of photosynthesis, which brought about a radical alteration in the level of atmospheric oxygen. From that invention had come an explosion of life forms, complex multicellular organisms creeping to the surface of the sea and eventually crawling out to colonize the land. By the mid-Paleozoic period oxygen emanating from these organisms had increased to 20 percent of the atmosphere. Although it dropped precipitously in the late Paleozoic, eventually the oxygen returned to that same 20 percent level and, over the past 100 million years, has remained there in an "oscillating steady state." Odum warned, however, that "man-generated CO_2 and dust pollution might be making this precarious balance still more 'unsteady.' "[5]

I would call this changing global picture of organisms interacting with the atmosphere a kind of history and, on the basis of this picture, would include Professor Odum among the ranks of historians, broadly defined. But what a strange history he tells when compared to the history we humans make. If in the 1960s I had written a history of the United States that described the nation as moving through a series of predictable "developmental stages" to a condition of maturity characterized by lower net production but higher stability (that is, resistance to external perturbations), higher diversity, closed mineral cycles, good nutrient conservation, and low entropy, my colleagues would have wondered what substance I was abusing. Unlike nature, the nations of the world, it was commonly understood, may "develop," but they never reach a steady state. In the 1960s the United States was certainly a highly developed country, at least in industrial terms, but its population was growing, not stabilizing, its resources were dwindling, its cities were burning, its streets were filled with antiwar protesters, several of its leaders were getting shot. An observer of those changes might have asked why the history of American society should be so much more chaotic than the history of an oak/hickory forest. Or why should the past thousand years of human activity look so much more unsettled than the past hundred million years of other species? Odum did not reflect on the contrast, but he did suggest that the major agency threatening to upset the global balance at this point is a single runaway species burning fossil fuels with abandon and stirring up the dirt.

History, even very recently, was still compartmentalized into two separate spheres, one for people and one for the rest of the natural world. The first, the human story, had been tumultuous, unpredictable, and destructive; the second had been, with big exceptions on the largest scale of time, orderly, predictable, and conserving. The great challenge facing us, according to the popular ecological literature of the 1960s, was to bring our human history into conformity with the history of nature and thereby achieve a unified steady state.

I grew up with that sort of thinking, as did many other environmental historians as well as ecologists and conservationists, and believed that we must try to reintegrate a degrading human life into the more orderly patterns of nature. There still seems to be good evidence and solid reason to support that sort of thinking. The history we are writing on the planet has become more destructive than ever, destructive of species, communities, ecosystems, and our own security, and it clearly needs a different model of change. But can nature furnish that model? Does the history of nature, as we understand it today, still furnish any compelling norms for the history of humankind?

If ecology was an inspiring guide a few decades back, it is no longer so today in the eyes of many observers, including many ecologists. Over the past two decades the field of ecology has pretty well demolished Eugene Odum's portrayal of a world of ecosystems tending toward equilibrium, leaving us with no model of development for human society to emulate. I have written about that demolition elsewhere.[6] It begins with an increasing interest in the phenomenon of environmental disturbance. "Disturbance" was not a common subject in Odum's heyday, and it almost never appeared in combination with the adjective "natural." By the 1970s, however, scientists were out looking strenuously for signs of disturbance in nature—especially signs of disturbance that were not caused by humans—and they were finding them everywhere. By the present decade revisionistic ecologists have succeeded in leaving little tranquility in primitive nature. Fire is one of the most common disturbances they have noted. So is wind, especially in the form of violent hurricanes and tornadoes. So are invading populations of microorganisms and pests and predators. And volcanic eruptions. And invading ice sheets of the Quaternary period. And devastating droughts like that of the 1930s in the American West. Above all, it is this last category of disturbances, caused by the restlessness of climate, that the new generation of ecologists has emphasized. As one of the most influential of them, Professor Margaret Davis of the University of Minnesota, has written: "For the last 50 years or 500 or 1,000—as long as anyone would claim for 'ecological time'—there has never been an interval when temperature was in a steady state with symmetrical fluctuations about a mean. . . . Only on the longest time scale, 100,000 years, is there a tendency toward cyclical variation, and the cycles are asymmetrical, with a mean much different from today."[7]

A representative expression of the new post-Odum ecology is a book of essays edited by S.T.A. Pickett and P. S. White, *The Ecology of Natural Disturbance and Patch Dynamics*, published in 1985. I offer it as symptomatic of much of the thinking going on today in the ecology field. Though the final section of the book does deal with ecosystems, the word has lost much of its former implication of order and equilibrium. Two of the authors in fact open their contribution with a complaint that many scientists assume that "homogeneous ecosystems are a reality," when in truth "virtually all naturally occurring and man-disturbed ecosystems are mosaics of environmental conditions." "Historically," they write, "ecologists have been slow to recognize the importance of disturbances and the heterogeneity they generate." The reasons for this reluctance have to do with cultural assumptions rooted in the past: "The majority of both theoretical and empirical work has been dominated by an equilibrium perspective."[8] Repudiating that perspective and embracing a more modern one of constant change, these authors take us to the tropical forests of South and Central America and to the Everglades of Florida, showing us instability on every hand: a wet, green world of continual disturbance—or, as they prefer to say, "of perturbations." Even the grasslands of North America, which inspired Frederic Clements' theory of the climax, appear in this collection as regularly disturbed environments. One paper describes them as a "dynamic, fine-textured mosaic" that is constantly kept in upheaval by the workings of badgers, pocket gophers, and mound-building ants, along with fire, drought, and eroding wind and water.[9] The message in all these papers is consistent: the old ideal of equilibrium is dead;

the ecosystem has receded in usefulness; and in their place we have the idea of the lowly "patch." Nature should be regarded as a landscape of patches of all sizes, textures, and colors, changing continually through time and space, responding to an unceasing barrage of perturbations.

Now this is a nature which looks remarkably similar to the human community that academic departments of history have been writing about. We have identified our own patches; they have names like the United States (which in turn is a patchwork quilt of racial, ethnic, and gender differences) or California or the Salinas Valley. We can no longer pretend that any of them is a stable entity: history reveals that they often shift in actual physical dimensions—the United States of today is radically larger than it was two hundred years ago, the Soviet Union has dissolved almost overnight—and they often shift in their internal composition—California's population not only growing dramatically in size but also changing dramatically in racial composition. Disturbance comes from a congeries of cultural and natural agents, including droughts, earthquakes, pests, viruses, corporate invasions, loss of markets, new inventions, crimes, federal laws, and even French literary theory. Disturbance *is* history. And a disturbed nature is a nature that has a history very similar to the history that humans make.

Like ecologists, historians of social change once had a fairly clear ideal of what a normal society must look like. White men must be in charge, it was assumed, presiding over their families with dignity and over the affairs of state with gravity. They work to achieve progress year by year, but that progress does not upset the social equilibrium. They come up with many interesting new ideas, but they go on worshiping the same God. They follow reason and revelation together as their guides. They extend the blessings of civilization to less fortunate peoples of the world. If a war breaks out, then people must fight it; but afterward they try to restore the world to normalcy. That norm, many historians believed as did as the general public, was a lot like Queen Victoria's Great Britain, or like Warren G. Harding's home in small-town Ohio. Thus people lived in history without (they thought) letting it overwhelm them. But the last twenty years have been hard on all norms, and they have left historians as confused as anyone else about what is normal or desirable in the midst of change.

One of the most important insights that the discovery of history has brought us to see is that all ideas, past and present, are grounded in particular historical contexts. This generalization includes the ideas of politicians, businesspeople, scientists, and even historians—it covers *all* ideas. We call this insight the principle of historicism: the theory that historians must avoid all value judgments in studying past periods or former cultures; in other words, a theory of historical relativism. It is supposed to give us greater objectivity toward and sympathy for the people of the past who could not share our blessings of enlightenment; at the same time, it is supposed to free us from any blind allegiance to present-day opinions. If we must explain the past in its own terms, as historicism argues, then we must also be wary of uncritically accepting the terms that condition our own way of thinking.

Carried far enough, the philosophy of historical relativism teaches us that we must even try to free ourselves from current value judgments and write dispassionately the history of our obsession with history (an obsession that has been growing since the early nineteenth century), and we must also try to understand the fixation on disorder that has characterized our own time. We must *explain* this newest obsession. Where is this emphasis on change and disorder coming from?

The obvious answer is that it is coming from the experience of rapid social change that has been accelerating over the past two centuries. Earlier generations, going back hundreds of thousands of years, experienced change too, but in a radically different context. According to Claude Lévi-Strauss, "the characteristic feature of the savage mind is its timelessness; its object is to grasp the world as both a synchronic and a diachronic totality."[10] In post–hunting and gathering societies, when agriculture began to dominate daily life, the idea of change remained more cyclical than linear; the recurrent cycle of annual crops was more immediately real to people than the long-term evolution of human life. Nature appeared as a stable order, created in the beginning of time by decree or flung spontaneously into being and never altered in its essential properties or relations. But this is not the way modern people understand the world or time, and the reason lies in changing material and cultural circumstances.

We live on the other side of a revolution in consciousness brought about by the explosive forces of modern capital, technology, and materialism. The description of these forces is too complicated to go into here, but despite a bit of hyperbole and oversimplification Karl Marx was right when he credited the economic culture associated with modern capitalism with creating a new passion for change and a new attitude toward time:

> Constant revolutionizing of production, uninterrupted disturbance of all social conditions, everlasting uncertainty and agitation distinguish [this] epoch from all earlier ones. All fixed, fast-frozen relations, with their train of ancient and venerable prejudices and opinions, are swept away. All new-formed ones become antiquated before they can ossify. All that is solid melts into air, all that is holy is profaned.[11]

Marx was thinking primarily about the effects of capitalism on ideas of social community, as people shifted from a rural to an urban setting, but we can see how readily his words also apply to ideas of the natural order. The sense of the ecological whole that once seemed so solid and unshakable has tended, along with all other ideas, to melt into air.

Karl Marx welcomed this new time consciousness, indeed, built his theory of dialectical materialism on it, following the great philosopher of history Georg Wilhelm Friedrich Hegel. Marx believed that the destruction of traditional ideas was necessary to free people from the prejudices of the past. You cannot find in him or his nineteenth-century disciples, therefore, much concern about preserving any traditional feeling for nature or even any concern for environmental preservation. But the advocates of socialism did predict that one day history would come to an end in a timeless utopia of the classless society. The disorder of economic revolution would cease and society would finally reach a steady state of established relations, an equilibrium of justice, from which it would follow that nature would also arrive at some point of equilibrium, though one that was firmly under human control. That prediction now seems to have been proved wrong. A century after Marx's death the tumult of change has not ceased, indeed, has not even slowed. Nor have we approached the ideal of justice for all; on the contrary, we have seen the exacerbation of global inequalities, not their disappearance. Today, for many former believers, the socialist dream of a glorious end to capitalist history has collapsed and lies in shards.[12]

Industrial capitalism, blaring its triumph over all challengers, promising a "new world order" of the endless accumulation of wealth, makes no promise of ever achieving a steady state in social, economic, or ecological terms. Its vision remains one of ceaseless change, infinite possibilities, and boundless creativity. In light of its past record we can expect that global capitalism will continue to promote unchecked economic and population growth, will continue to stoke the rising aspirations of the poor, and will intensify its currently intense demands on nature. The effect of these explosive forces will be to dissolve any fragmentary notions of stability, order, or normality left to us, and will leave us more than ever dwelling in a world where change has become the very purpose of life.

So, in this manner, we historians can make a reasoned explanation of the modern tendency to turn nature into a mirror of our society, reflecting back the chaotic energies of capital and technology. And by making that explanation we can free ourselves from mindless, uncritical allegiance to the present orthodoxy, as historical analysis has liberated us from previous orthodoxies and promoted critical thinking. Fortified by the principle of historical relativism we can approach these recent ecological models that dramatize disturbance with a sense of freedom and independence. If they are not the mere reflection of global capitalism and its ideology, they are nonetheless highly compatible with that force dominating the earth. The newest ecology, with its emphasis on competition and disturbance, is clearly another manifestation of what Frederic Jameson has called the "logic of late capitalism."[13] The most recent models of ecology are sure to be superseded, we can assume, just as Odum's model of the ecosystem was superseded. And realizing that fact, we can conclude they are not necessarily more truthful than their predecessors or their successors.

But having glimpsed this connection between the science of ecology and its cultural and economic conditions, where are we? Are we then free to believe anything else instead? The answer must be yes, and yet also no. The theory of historical relativism frees us from dogma but offers no firm guidance to belief. It cannot really invalidate the intellectual tendencies of our time, or any other time, nor can it validate new ones. On the contrary, it can only lead either to complete cynicism or to the acceptance of any set of ideas or any environment that humans have created as legitimate. Disneyland, by the theory of historical relativism, is as legitimate as Yellowstone National Park, a wheat field is as legitimate as a prairie, a megalopolis of thirty million people is as legitimate as a village. Each is the product of history and therefore stands equal to its opposite. Each has its own logic to be penetrated and understood, but any logic, like any set of beliefs or institutions appearing over time, must appear to the consistent relativist to be as good as any other.

If the study of human or natural history forced us to adopt such a rigidly relativist position, then I would be ready to join those who call for the wholesale rejection of modern historical consciousness as a degenerate worldview.[14] I would be seeking some way back to a prehistorical or premodern consciousness—to some folk or vernacular worldview that preceded modern historical thinking. But such a conclusion is not required of us; the practice of history does not require so radical a version of historical relativism. We can pursue a knowledge of the past, going back two million years for humans, billions for the rest of nature, without getting completely lost in the labyrinths of time. I want to suggest now several conclusions that can inform our search for solid values. They are conclusions that transcend our present-day circumstances. They seem to be objectively true, too, supported by substantial evidence drawn from the intertwined study of nature and humanity. They are conclusions about nature and about human society, acknowledging that we cannot set up any impermeable barrier between the two realms. And they are conclusions about reality that not only can but must constrain our choices and inform our values if we want to survive.

In the first place, modern knowledge reveals that living nature, for all its private, individualistic strivings, works by the principle of interdependency. Indeed, it can work only by that principle; no species, plant or animal, no person in society, has any chance of surviving without the energy or aid of others. New proof of the interdependency principle has come to light in these days of extraterrestrial travel: send any individual organism into outer space alone, without any of the services provided by other kinds of organisms, from soil fertility to oxygen generation, and it will not survive. Sooner or later it will need its evolutionary companions. This is not simply the discovery of today's natural science, however; it was also the discovery of the earliest cultures we know about. Knowledge of the manifold forms of this interdependency has been accumulating since Lucy walked across the plains of Ethiopia. All the changes we can find in civilization, it is now clear, are only changes in the patterns of this interdependency, not in the reality or necessity of interdependency itself. Even individualistic America, despite its constant extolling of self-reliance, is a society that literally lives by the practice of mutual aid, from the production of food to the provision of defense, education, health care, and personal security, and in some measure it always has done so.

To be sure, for a while many people lost sight of that truth, even began to imagine they could live by technological innovations alone. But the past few decades have demolished that illusion. What we call the environmental movement of the post–World War Two period has been essentially a reawakening to the truth, grounded in experience, that we must depend on other forms of life to survive; we have no options. Progress has not made our condition different in this respect from the condition of our most remote ancestors. Being clever and adaptable, we have learned how to make substitutions in our dependencies and to alter the geography of our dependency—for example, we can learn to raise and eat Central American beef instead of Canadian moose—but we have not learned how to live on a planet that is dead or devoid of others.

The full implications of our ecological dependency are still working their way into the heads of economic and political leaders, but already these implications are eroding any grandiose claims of conquest over the earth or boasts of our invulnerability before the forces of nature. Consequently, the extinction of obscure species has become a global concern expressed in international treaties. Communities, states, and nations are no longer so sure they can manage without these species, even if many of them play only a remote, distant, or obscure role in human welfare. At the same time an awareness of our dependency on the whole fabric of life is stimulating a sense of dependency on other people, most of them strangers to us but locked with us in a common predicament. Again, the forms of dependency may change. The solidarity of the face-to-face group, working together for survival, may become transformed into something larger and perhaps less effective—into a single global audience instantly tuned in to the fate of victims of disease, tyranny, poverty, or forest destruction wherever they live. Thus the fact of interdependency binding all living things into a kind of community has not been invalidated by the rapid pace of recent change or the many uncertainties that change has produced.

In the second place, our study of the past has revealed models of successful adaptation that we can learn from today and use to inform our values. They are not values in themselves but, rather, lessons drawn from nature. The natural world may not provide any overall, sufficient norm for us to follow, or any single transcendent good that we can discover, but it does provide a wealth of models, depending on what it is we want to achieve. If we want to fly, for example, we can find models in the wings of birds, models that took tens of millions of years to perfect. If we want to stop soil erosion or survive drought, we have a model in the tallgrass prairie, which retains far more of the rain that falls than a wheat monoculture does and can bound back from a severe dry spell that would wipe out a planted domestic crop. We may not think about such models as lessons derived from history, but they are all the products of past experience, and it is the biologist, thinking historically, who reveals how they came into being, by a process of evolution that we can call the unfolding wisdom of life.

Similarly, environmental history sets before us models of human communities that have been more successful than we have in some respects. For example, if social longevity is high in our hierarchy of values, if we want to survive as a people and as a species for the longest possible time, then we can find in the past a wealth of examples that have something useful to teach. We cannot find in the past or present any societies that are perfect in every aspect, or examples that we can simply revive lock-stock-and-barrel from extinction; but we can find models to study and learn from. They exist within the borders of the United States and in every part of the earth—communities that have managed to fit themselves to their places for impressively long periods of time, that are less destructive of the biota around them, that may have acquired some vital knowledge of place which we lack. They may have not escaped the hand of time, but they have come closer than we have to adapting to it. My own research as a historian suggests that such enduring communities have had one dominant characteristic: they have made rules, and many of them, rules based on intimate local experience, to govern their behavior. They have not tried to "live free" of nature or of the group; nor have they resented restraints on individual initiative or left it to each individual to decide completely how to behave. On the contrary, they have accepted many kinds of limits on themselves and enforced them on one another. Their methods of enforcement may not meet our modern American standards of privacy or justice; they may not be compatible with our modern sense of strong personal rights; and certainly they can stifle creativity or originality. But throughout history, having these rules and enforcing them vigorously seems to be a requirement for long-term ecological survival.

How we translate such models from other eras and places into values for our own lives is a very difficult question. Clearly, we cannot merely turn all our wheat fields, no matter how inadequate ecologically they may be, back into bluestem prairies, nor can we turn industrial capitalism back into a medieval alpine village or an Australian bushmen's camp. We simply cannot go back in time and undo all that has happened. We are, in that sense, prisoners of time. But we can approach the record of the past with much more respect, admitting that most of the innovations we have recently made are not likely to survive, that what is old among us is by that very fact worthy of respect and mimicry, that what is *very* old is likely to be wise.

In the third place, history reveals not merely that change is real but also that change is various. All change is not the same, nor are all changes equal. Some change is cyclical, some is not. Some changes are linear, some are not. Some changes take an afternoon to accomplish, some a millennium. We can no more take any particular kind of change as absolutely normative than we can take any particular state of equilibrium as normative. The fact that ice sheets once scraped their way across Illinois does not provide any kind of justification for a corporation that wants to strip coal from the state. We know this, but sometimes we get confused by talk about all change being "natural." In a loose sense, the statement is true; but it is also meaningless. No one really maintains that whatever is, is right, or that whatever happens is good. We understand that there are changes in nature that work against us as well as for us, changes that we have to defend ourselves against, even if we cannot prevent them. The challenge is to determine which changes are in our more enlightened self-interest and consistent with our most rigorous ethical reasoning, always remembering our inescapable dependence on other forms of life.

Environmental conservation becomes, by this way of historical thinking, an effort to protect certain rates of change going on within the biological world from incompatible changes going on within our economy and technology. It is not a program of locking nature up within a museum case, freezing it for all time. Rather, it is an effort based on the idea that preserving a diversity of change ought to stand high in our system of values—that promoting the coexistence of many beings and many kinds of change is a rational thing to do. The pace of innovation in computer chips may be appropriate to a competitive business community, but it is not appropriate to the evolution of a redwood forest. Some things take longer to grow or improve. Some things cannot adapt as fast as others. These are differences revealed by history. Today, historians can no longer claim there is a single universal narrative of change that all species, all communities, all places, must conform to. "History" has given way to "histories." Each of these histories needs space to play itself out, to unwind its narrative. This is precisely what the modern idea of conservation aims to do: provide the space, either set aside in large, discrete blocks or protected within the interstices of the landscape, so that all the many histories can coexist—the history of a coral reef alongside the history of a coastal city, the history of a tropical rainforest alongside the history of a political struggle. This strategy of trying to conserve a diversity of changes may seem paradoxical, but it is founded on a crucial and reasonable insight. We may live by change, and we may be the products of change, but we do not always know—indeed, we cannot always know—which changes are vital and which are deadly.

These are conclusions about the real world, I believe, that the study of nature and history leads us to make today. They are conclusions that stand up well because they are based not merely on private fantasies but on knowledge. They make sense of the scene outside my window. A hundred years from now they will likely be as valid as they are today, though that scene may become very different: it may become a landscape desiccated by global warming, it may be uprooted for a housing development, or it may even be emptied of all birds and trees. History tells plenty of stories of human folly, of death and decline. Whether we choose to learn from the past or not, whatever we choose to learn or ignore, the past is our only instructor. We no longer have nature in some timeless state of perfection, nor revelation nor authority, to depend on. From that changing past, and from it only, we must somehow draw, with the aid of imperfect reason, what we value and defend.

Notes

1. This is the argument made by Calvin Luther Martin in his eloquent book *In the Spirit of the Earth: Rethinking History and Time* (Baltimore: Johns Hopkins University Press, 1992), which sees modern history as a force corrupting the ancient hunter-gatherer relation with nature. For Martin, history is so deeply imbued with the ideology of progress as to be identified with it, an identification that many historians would dispute. They point out that Martin is himself still a historian, though one who tells a story of decline rather than progress.

2. Stephen Toulmin and June Goodfield, *The Discovery of Time* (Chicago: University of Chicago Press, 1965), p. 17.

3. Ibid., p. 232.

4. Eugene P. Odum, *Fundamentals of Ecology*, 3rd ed. (Philadelphia: Saunders, 1971), pp. 251–252.

5. Ibid., pp. 271–272. This discussion forms the core of the Gaia hypothesis, developed by James Lovelock; see, for example, his *Ages of Gaia: A Biography of Our Living Earth* (Oxford: Oxford University Press, 1988).

6. See my essay, "The Ecology of Order and Chaos," in *The Wealth of Nature: Environmental History and the Ecological Imagination* (New York: Oxford University Press, 1993), pp. 156–170.

7. Margaret Bryan Davis, "Climatic Instability, Time Lags, and Community Disequilibrium," in *Community Ecology*, edited by Jared Diamond and Ted J. Case (New York: Harper & Row, 1986), p. 269.

8. James R. Karr and Kathryn E. Freemark, "Disturbance and Vertebrates: An Integrative Perspective," in *The Ecology of Natural Disturbance and Patch Dynamics*, edited by S.T.A. Pickett and P. S. White (Orlando: Academic Press, 1985), p. 154–155. The Odum school of thought is, however, by no means silent. Another recent compilation has been put together in his honor, and many of its authors express continuing support for his ideas: L. R. Pomeroy and J. J. Alberts (eds.), *Concepts of Ecosystem Ecology: A Comparative View* (New York: Springer-Verlag, 1988).

9. Orie L. Loucks, Mary L. Plumb-Mentjes, and Deborah Rogers, "Gap Processes and Large-Scale Disturbances in Sand Prairies," in Pomeroy and Alberts, *Concepts of Ecosystem Ecology*, pp. 72–85.

10. Claude Lévi-Strauss, *The Savage Mind* (Chicago: University of Chicago Press, 1966), p. 263.

11. Karl Marx and Fredrich Engels, "Manifesto of the Communist Party," in *Basic Writings on Politics and Philosophy: Karl Marx and Fredrich Engels*, edited by Lewis S. Feuer (Garden City, N.Y.: Anchor Books, 1959), p. 10.

12. A new socialism, to be sure, has emerged in recent years—a "green" or "eco" socialism that seeks to recover the neglected insights of Karl Marx into the significance of the human/nature relation and to link the cause of protecting the earth with that of achieving a more equitable distribution of resources. The journal *Capitalism Nature Socialism*, edited by James O'Connor, is the best guide to this movement.

13. See Frederic Jameson, *Postmodernism: The Cultural Logic of Late Capitalism* (Durham: Duke University Press, 1991). For another, more positive, view of postmodernism and its relation to capital, see David Harvey, *The Condition of Postmodernity: An Enquiry into the Origins of Cultural Change* (Oxford: Blackwell, 1989). Harvey believes that we are freeing ourselves not only from the logic of capital but also from the authority of objective science that has so long dominated our consciousness. If this is true, then nature must become little more than a private vision and lose all claim to serving as a norm or guide in any degree for humanity.

14. The real threat of modern historical consciousness lies not in its embrace of the ideology of progress, which assumes that the present state of affairs is not only inescapable but good. Rather, I am suggesting, it lies in its tendency to embrace a theory of total relativism.

CHAPTER SIX

CULTURAL PARALLAX IN VIEWING NORTH AMERICAN HABITATS

GARY PAUL NABHAN

A debate is raging with regard to the "nature" of the North American continent—in particular, the extent to which habitats have been managed, diversified, or degraded over the last ten thousand years of human occupation (Gomez-Pompa and Kaus 1992). This debate has at its heart three issues: whether the "natural condition of the land" by definition excludes human management; whether officially designated wilderness areas in the United States should be free of hunting, gathering, and vegetation management by Native Americans or other people; and whether traditional management by indigenous peoples is any more "benign" or "ecologically sensitive" than that imposed by resource managers trained in the use of modern Western scientific principles, methods, and technologies.

The debate, then, is not about the human mental "construction" of nature so much as it is about the physical "reconstructions" of habitats by humans and to what extent these are perceived as "natural" or "ecological." The debate is not merely an academic dispute. It involves hunters, gatherers, ranchers, farmers, and political activists from a variety of cultures, not just "Western scientists" versus "indigenous scientists." The outcome will no doubt shape the destiny of officially designated wilderness areas in national parks and forests throughout North America (Gomez-Pompa and Kaus 1992; Flores et al. 1990).

Consider, for example, the declaration of the 1963 Leopold Report to the U.S. Secretary of the Interior: that each large national park should maintain or recreate "a vignette of primitive America," seeking to restore "conditions as they prevailed when the area was first visited by the white man"—as if those conditions were synonymous with "pristine" or "untrammeled" wilderness (Anderson and Nabhan 1991:27). Such a declaration either implies that pre-Columbian Native Americans had no impact on the areas now found within the U.S. National Park System or that indigenous management of vegetation and wildlife as it was done in pre-Columbian times is compatible with and essential to "wilderness quality." For Native Americans with historic ties to land, water, and biota within parks, this latter interpretation provides them a platform for being *co-managers*, not merely harvesters of certain traditionally utilized resources, as currently sanctioned by the National Park Service (1987).

On one side of the debate are those who argue that Native Americans have had a negligible impact on their homelands and left large areas untouched. That is to say, these original human inhabitants did little to actively manage or influence wildlife populations one way or another. An early proponent of this view was John Muir: "Indians walked softly and hurt the landscape hardly more than the birds and squirrels, and their brush and bark huts last hardly longer than those of wood rats, while their enduring monuments, excepting those wrought on the forests by fires they made to improve their hunting grounds, vanish in a few centuries." Yet the Yosemite landscapes he knew so well are now known to have been dramatically shaped by Native American management practices (Anderson and Nabhan 1991:27).

Some proponents of this perspective even deny Muir's exception that controlled burns had a significant impact. Native Americans, they claim, would have no interest in managing the forests even if they were capable of it (Clar 1959:7): "It would be difficult to find a reason why the Indians [of California] should care one way or another if the forest burned. It is quite something else again to contend that the Indians used fire systematically to 'improve' the forest. Improve it for what purpose? . . . Yet this fantastic idea has been and still is put forth time and again."

A second stance on this side of the debate contends that Native American spirituality kept all members of indigenous communities from harming habitats or the biota within them. Leslie Silko (1986:86), who is of Laguna Pueblo descent, has argued that "survival depended upon the harmony and cooperation, not only among human beings, but among all things—the animate and inanimate. . . . As long as good family relations [between all beings] are maintained . . . the Earth's children will survive." The implication is that Native Americans practiced a spirituality "earthly enough" to restrain any tendencies toward overharvesting or toward depletion of diversity through homogenizing habitat mosaics (Anderson 1993). As Max Oelschlaeger (1991:17) has assumed, *"harmony with rather than exploitation of the natural world was a guiding principle for the Paleolithic mind and remains a cardinal commitment among modern aborigines."

The other side of this argument contends that Native Americans and other indigenous peoples have rapaciously exterminated wildlife within their reach and that their farming, hunting, and gathering techniques were often ecologically ill suited for the habitats in which they were practiced. Kent Redford (1985) and others have taken such a stance to play devil's advocate with the romantic notion of "the ecological noble savage." Award-winning science writer Jared Diamond (1993:268) has also tried to dispel what he sees as a myth of native peoples as "environmentally minded paragons of conservation, living in a Golden Age of harmony with nature, in which living things were revered, harvested only as needed, and carefully monitored to avoid depletion of breeding stocks."

Diamond (1993:263, 268) claims that in thirty years of visiting native peoples on the three islands of New Guinea, he has failed to come across a single example of indigenous New Guineans showing friendly responses to wild animals or consciously managing habitats to enhance wildlife populations: "New Guineans kill those animals that their technology permits them to kill," inevitably depleting or exterminating more susceptible species. His claim that all indigenous New Guinean cultures respond in the same manner to nature is astonishing when one considers that about one thousand of the world's remaining languages are spoken in New Guinea and that this cultural/linguistic diversity would presumably encode many distinctive cultural responses to the flora and fauna.

Yet to my knowledge, Diamond himself has never objectively field-tested his game depletion hypothesis as Vickers (1988) has done among Amazonian peoples, where it was demonstrated over several years that native hunters would switch to less desirable prey before locally extirpating rare game species. Redford and Robinson (1987) have also documented that indigenous South American hunters take a wider variety of game species than neighboring nonindigenous South American colonists, who are more likely to use degraded habitats near larger settlements, thereby further impoverishing the abundance and diversity of wildlife. Diamond has never demonstrated that he has systematically asked indigenous consultants about less obvious techniques by which hunters and gatherers may influence habitat quality and wildlife abundance, following research protocols such as those outlined by Blackburn and Anderson (1993:24). And yet, repeatedly, Diamond (1986, 1992, and elsewhere) has used the hypotheses of the Pleistocene overkill, and later selective cutting of fir and spruce in Chaco Canyon by Anasazi city-state dwellers, to indict all Native American hunters, gatherers, and farmers as exterminators of wildlife and aggravators of soil erosion.

This dismissal of enormous historic and cultural differences is at the heart of the problem inherent in most discussions of "the American Indian view of nature" and assessments of the pre-Columbian condition of North American habitats. To assume that even the Hopi and their Navajo neighbors think of, speak of, and treat nature in the same manner is simply wrong. Yet individuals from two hundred different language groups from three historically and culturally distinct colonizations of the continent are commonly lumped under the catchall terms "American Indian" or "Native American."

Even within one mutually intelligible language group, such as the Piman-speaking O'odham, there are considerable differences in what taboos they honor with respect to dangerous or symbolically powerful animals. While the River Pima do not allow themselves to eat badgers, bears, quail, or certain reptiles for fear of "staying sickness," these taboos are relaxed or even dismissed by other Piman groups who live in more marginal habitats where game is less abundant (Rea

1981; Nabhan and St. Antoine 1993). An animal such as the black bear—which is never eaten by one Piman community because it is still considered to be one of the "people"—is routinely hunted by another Piman-speaking group which prizes its skin and pit-roasts its meat—an act that would be regarded much like cannibalism in the former group (Rea 1981; personal communication). Moreover, contemporary Pima families do not necessarily adhere to all the taboos that were formerly paramount to all other cultural rules which granted someone Piman identity.

Despite such diversity within and between North American cultures, it is still quite common to read statements implying a uniform "American Indian view of nature"—as if all the diverse cultural relations with particular habitats on the continent can be swept under one all-encompassing rug. Whether one is prejudiced toward the notion of Native Americans as extirpators of species or assumes that most have been negligible or respectful harvesters, there is a shared assumption that all Native Americans have viewed and used the flora and fauna in the same ways. This assumption is both erroneous and counterproductive in that it undermines any respect for the realities of cultural diversity. And yet it continues to permeate land-use policies, environmental philosophies, and even park management plans. It does not grant *any* cultures—indigenous or otherwise—the capacity to evolve, to diverge from one another, or to learn about their local environments through time.

This distortion of the relationships between human cultures and the rest of the natural world is what I call "cultural parallax of the wilderness concept." If you remember your photography or astronomy lessons, *parallax* is the apparent displacement of an observed object due to the difference between two points of view. For example, consider the difference between the view of an object as seen through a camera lens and the view through a separate viewfinder. A cultural parallax, then, might be considered to be the difference in views between those who are actively participating in the dynamics of the habitats within their home range and those who view those habitats as "landscapes" from the outside. As Leslie Silko (1987:84) has suggested: "So long as human consciousness remains *within* the hills, canyons, cliffs, and the plants, clouds, and sky, the term *landscape*, as it has entered the English language, is misleading. 'A portion of terri-

tory the eye can comprehend in a single view' does not correctly de-
scribe the relationship between a human being and his or her surroundings."

Adherents of the romantic notion of landscape claim that the most
pristine and therefore most favorable condition of the American continent worthy of reconstruction is that which prevailed at the moment of European colonization. As William Denevan (1992) has
amply documented, the continent was perhaps most intensively
managed by Native Americans for the several centuries prior to Columbus's arrival in the West Indies. Because European diseases decimated native populations through the Americas over the following
hundred and fifty years, the early European colonists saw only vestiges of these managed habitats, if they recognized them as managed
at all (Cronon 1983; Denevan 1992).

And yet, among many ecologists, including Daniel Botkin
(1990:195), "the idea is to create natural areas that appear as they did
when first viewed by European explorers. In the Americas, this would
be the landscape of the seventeenth century. . . . If natural means
simply *before human intervention*, then all these habitats could be
claimed as natural." Thus Botkin equates the periods prior to European colonization with those prior to *human intervention* in the landscape and assumes that all habitats were equally pristine at that time.
By this logic, either the pre-Columbian inhabitants of North America were not human or they did not significantly interact with the
biota of the areas where they resided.

Human influences on North American habitats began at least
9,200 years prior to the period Botkin pinpoints—when newly arrived "colonists" came down from the Bering Strait into ice-free
country (Janzen and Martin 1982; Martin 1986). Regardless of how
major a role humans played in the Pleistocene extinctions, the loss of
73 percent of the North American genera of terrestrial mammals
weighing one hundred pounds or more precipitated major changes in
vegetation and wildlife abundance. By Paul Martin's (1986) criteria,
North American wilderness areas have been lacking "completeness"
for over ten millennia and would require the introduction of large
herbivores from other continents to simulate the "natural conditions" comparable to those under which vegetation cover evolved
over the hundreds of thousands of years prior to these extinctions.

It has always amazed me that many of the same scholars who are willing to grant pre-Columbian cultures of the Americas more eco-logical wisdom than recent European colonists still deny the possi-bility that these cultures could have played a role in these faunal ex-tinctions, as if that wisdom did not take centuries to accumulate. Do they believe that the pre-Columbian cultures of North America be-came "instant natives" incapable of overtaxing any resources in their newfound homeland—an incapability that few European cultures have achieved since arriving in the Americas five centuries ago? As Michael Soulé (1991:746) has pointed out, "the most destructive cul-tures, environmentally, appear to be those that are colonizing unin-habited territory and those that are in a stage of rapid cultural (often technological) transition." My point is simply this: it may take time for any culture to become truly "native," if that term is to imply any sensitivity to the ecological constraints of its home ground.

I am not arguing that many indigenous American cultures did not develop increasing sensitivity to the plant and animal populations most vulnerable to depletion within their home ranges. To the con-trary, I would like to suggest that all of pre-Columbian North Amer-ica was not pristine wilderness for the very reason that many indige-nous cultures actively managed habitats and plant populations within their home ranges as a response to earlier episodes of overexploita-tion. There is now abundant evidence that hundreds of thousands of acres in various bioregions of North America were actively managed by indigenous cultures (Anderson and Nabhan 1991; Denevan 1992; Fish et al. 1985). This does not mean that the entire continent was a Garden of Eden cultivated by Native Americans, as Hecht and Posey have erroneously implied for the Amazon (Parker 1992). Many large areas of the North American continent remained beyond the influ-ence of human cultures, and should remain so. Nevertheless, it is clear that the degree to which North American plant populations were consciously managed—and conserved—by local cultural tradi-tions has been routinely underestimated.

Hohokam farmers, for example, constructed over seventeen hun-dred miles of prehistoric irrigation canals along the Salt River in the Phoenix basin and intensively cultivated and irrigated floodplains along the Santa Cruz and Gila rivers, as well as on intermittent wa-tercourses for one hundred and fifty miles south and west of these pe-rennial streams (Nabhan 1989). At the same time, they cultivated agave relatives of the tequila plant over hundreds of square miles of upland slopes and terraces beyond where modern agricultural techniques allow crops to be cultivated today (Fish et al. 1985). Nevertheless, the discovery of native domesticated agaves being grown on a large scale in the Sonoran Desert has been made only within the last decade, despite more than a half century of intensive archaeological investigation in the region. Earlier archaeologists had simply never imagined that pre-Columbian cultures in North America could have cultivated perennial crops on such a scale away from riverine irrigation sources.

In the deserts of southern California, indigenous communities transplanted and managed palms for their fruits and fiber in artificial oases, some of them apparently beyond the "natural distribution" of the California fan palm. Control burns were part of their management of these habitats, and such deliberate use of fire created artificial savannas in regions as widely separated as the California Sierra and the Carolinas. In the Yosemite area, where John Muir claimed that "Indians walked softly and hurt the landscape hardly more than the birds and squirrels," Anderson's (1993) reconstructions of Miwok subsistence ecology demonstrate that the very habitat mosaic he attempted to preserve as wilderness was in fact the cumulative result of Miwok burning, pruning, and selective harvesting over the course of centuries.

So what Muir called wilderness, the Miwok called home; the parallax is apparent again. Is it not odd that after ten to fourteen thousand years of indigenous cultures making their homes in North America, Europeans moved in and hardly noticed that the place looked "lived-in"? There are perhaps two explanations for this failure. One response, as historians William Cronon (1983), William Denevan (1992), and Henry Dobyns (1983) have suggested, is that many previously managed landscapes had been left abandoned between the time when European-introduced diseases spread through the Americas and the time when Europeans actually set foot in second-growth forests, shrub-invaded savannas, or defaunated deserts. The second explanation is that Europeans were so intent on taking possession of these lands and developing them in their own manner that they hardly paid attention to signs that the land had already been managed on a different scale and level of intensity.

It was easier for Europeans to assume possession of a land they considered to be virgin or at least unworked and uninhabited by people of their equal. Columbus himself had set out to discover unspoiled lands where the seeds of Christianity—a faith that was being corrupted in Europe, he felt—could be transplanted. In 1502, well after his own men had unleashed European weeds, diseases, and weapons on the inhabitants of the Americas, Columbus wrote to Pope Alexander VI claiming that he had personally visited the Garden of Eden on his voyages to the New World.

Even those who have condemned Europeans for the effects of their ecological imperialism on indigenous American cultures too often frame their concern as "conquistadors raping a virgin land." As "subjects of rape," American lands and their resident human populations are simply reduced to the role of passive victims, incapable of any resilience or dynamic response to deal capably in any way with such invasions. And so we are often left hearing the truism, "Before the White Man came, North America was essentially a wilderness where the few Indian inhabitants lived in constant harmony with nature"— even though four to twelve million people speaking two hundred languages variously burned, pruned, hunted, hacked, cleared, irrigated, and planted in an astonishing diversity of habitats for centuries (Denevan 1992; Anderson and Nabhan 1991). And we are supposed to believe, as well, that they all lived in some static homeostasis with all the various plants and animals they encountered.

As Daniel Botkin (1990) has convincingly argued, few predator/prey or plant/animal relationships have maintained any long-term homeostasis even where humans are not present, let alone where they are. Although there is little evidence to indicate that indigenous cultures regionally extirpated any rare plants or small vertebrate species, there are intriguing signs that certain prehistoric populations depleted local firewood sources and certain slow-growing fiber plants such as yuccas (Minnis 1978). Because different cultures used native species at different intensities, and each species has a different growth rate and relative abundance, it is impossible to generalize about the conservation of all resources in all places. Nevertheless, numerous localized efforts to sustain or enhance the abundance of certain useful plants have been well documented (Anderson 1993; Anderson and Nabhan 1991). It remains unclear whether by favoring certain useful plants over others, plant diversity increased or decreased in particu-

lar areas. To my knowledge, no study adequately addresses indigenous peoples' local effects on biological *diversity* (as opposed to their
effects on the *abundance* of key resources).

It can no longer be denied that some cultures had specific conservation practices to sustain plant populations of economic or symbolic
importance to their communities. In the case of my O'odham neighbors in southern Arizona, there have been efforts to protect rare
plants from overharvesting near sacred sites, to transplant individuals
to more protected sites, and to conserve caches of seeds in caves to ensure future supplies (Nabhan 1989). Landscape photographer Mark
Klett (1990:73) has written that too often wilderness in the European
American tradition is "an entity defined by our absence [as if] the
landscape does best without our presence." I find in O'odham oral literature an interesting counterpoint to this notion. The O'odham
term for wildness, *doajkam*, is etymologically tied to terms for health,
wholeness, and liveliness (Mathiot 1973). While it seems wildness is
positively valued as an ideal by which to measure other conditions,
the O'odham also feel that certain plants, animals, and habitats "degenerate" if not properly cared for. Thus their failure to take care of
a horse or a crop may allow it to go feral, but this degenerated feral
state is different from being truly wild. Similarly, their lack of attention to O'odham fields, watersheds, and associated ceremonies may
keep the rains from providing sufficient moisture to sustain both wild
and cultivated species. Many O'odham express humility in the face of
unpredictable rains or game animals, but they still feel a measure of
responsibility in making good use of what does come their way.

In short, the O'odham elders I know best still behave as active participants in the desert without assuming that they are ultimately "in
control" of it. This, in essence, is the difference between participating in *untrammeled* wilderness (as defined by the U.S. Wilderness
Act) and attempting to tame lands through manipulative management. (A *trammel* is a device which shackles, hobbles, cages, or con-
fines an animal, breaking its spirit and capacity to roam.) What may
look like uninhabited wilderness to outsiders is a habitat in which the
O'odham actively participate. They do not define the desert as it was
derived from the Old French *desertus*, "a place abandoned or left
wasted." Their term for the desert, *tohono*, can be etymologically
understood as a "bright and shining place," and they have long called

themselves the Tohono O'odham: the people belonging to that place. They share that place with a variety of plants and animals, a broad range of which still inhabit their oral literature (Nabhan and St. Antoine 1993).

Within their Sonoran Desert homeland, many O'odham people still learn certain traditional land management scripts encoded in their own Piman language, which they then put into practice in particular settings, each in their own peculiar way. What concerns me is that their indigenous language is now being replaced by English. Their indigenous science of desert is being eclipsed by more frequent exposure to Western science. And their internal or etic sense of what it is to be O'odham is being replaced by the mass media's presentation of what it is to be (generically) an American Indian. While I will be among the first to admit that change is presumably inherent to all natural and cultural phenomena, I am not convinced that these three changes are necessarily desirable. In virtually every culture I know of on this continent, similar changes are occurring with blinding speed. Both nature and culture are being rapidly redefined, not so much by what we learn from our immediate surroundings as by what we learn through the airwaves.

Let me highlight what Sara St. Antoine and I recently learned while interviewing fifty-two children from four different cultures, all of them living in the Sonoran Desert (Nabhan and St. Antoine 1993). Essentially we learned that with regard to knowledge about the natural world, intergenerational differences within cultures are becoming as great as the gaps between cultures. While showing a booklet of drawings of *common* desert plants and animals to O'odham children and their grandparents, for example, we realized that the children knew only a third of the names for these desert organisms in their native language that their grandparents knew. With the loss of those names, we wonder how much culturally encoded knowledge is lost as well. With over half the two hundred native languages on this continent falling out of use at an accelerating rate, a great diversity of perspectives on the structure and value of nature are surely being lost. And culture-specific land management practices are being lost as well.

One driving force in this loss of knowledge about the natural world is that children today spend more time in classrooms and in front of the television than they do directly interacting with their natural surroundings. The vast majority of the children we interviewed are now gaining most of their knowledge about other organisms vicariously: 77 percent of the Mexican children, 61 percent of the Anglo children, 60 percent of the Yaqui children, and 35 percent of the O'odham children told us they had seen more animals on television and in the movies than they had personally seen in the wild.

An even more telling measure of the lack of primary contact with their immediate nonhuman surroundings is this: a significant portion of kids today have never gone off alone, away from human habitations, to spend more than a half hour by themselves in a "natural" setting. None of the six Yaqui children responded that they had; nor had 58 percent of the O'odham, 53 percent of the Anglos, and 71 percent of the Mexican children. We also found that many children today have never been involved in collecting, carrying around, or playing with the feathers, bones, butterflies, or stones they find near their homes. Of those interviewed, 60 percent of the Yaqui children, 46 percent of the Anglos, 44 percent of the Mexicans, and 35 percent of the O'odham had never gathered such natural treasures. Such a paucity of contact with the natural world would have been unimaginable even a century ago, but it will become the norm as more than 38 percent of the children born after the year 2000 are destined to live in cities with more than a million other inhabitants. While few cities are entirely devoid of open spaces, manufactured toys and prefabricated electronic images have rapidly replaced natural objects as common playthings.

However varied the views of the natural world held by the myriad ethnic groups which have inhabited this continent, many of them are now converging on a new view—not so much one of experienced participants dynamically involved with their local environment as one in which they too may feel as though they are outside the frame looking in. Because only a small percentage of humankind has any direct, daily engagement with other species of animals and plants in their habitats, we have arrived at a new era in which ecological illiteracy is the norm. I cannot help concluding that we will soon be losing the many ways in which cultural diversity may have formerly enriched the biological diversity of various habitats of this continent. I can only hope that our children will pay more attention to this warning from Mary Midgley (1978:246) than we have:

Man is not adapted to live in a mirror-lined box, generating his own electric light and sending for selected images from outside when he needs them. Darkness and bad smell are all that can come from that. We need a vast world, and it must be a world that does not need us; a world constantly capable of surprising us, a world we did not program, since only such a world is the proper object of wonder.

References

Anderson, M. Kathleen. "The Experimental Approach to the Assessment of the Potential Ecological Effects of Horticultural Practices by Indigenous Peoples on California Wildlands." Ph.D. dissertation, University of California, Berkeley, 1993.

Anderson, Kat, and Gary Paul Nabhan. "Gardeners in Eden." *Wilderness* 55(194) (1991):27–30.

Blackburn, Thomas C., and Kat Anderson. "Introduction: Making the Domesticated Environment." In Thomas C. Blackburn and Kat Anderson (eds.), *Before the Wilderness: Environment Management by Native Californians*. Menlo Park: Ballena Press, 1993.

Botkin, Daniel. *Discordant Harmonies*. Oxford: Oxford University Press, 1990.

Clar, C. R. *California Government and Forestry from Spanish Days Until the Creation of the Department of Natural Resources in 1927*. Sacramento: California Division of Forestry, 1959.

Cronon, William. *Changes in the Land: Indians, Colonists, and the Ecology of New England*. New York: Hill & Wang, 1983.

Denevan, William M. "The Pristine Myth: The Landscape of the Americas in 1492." *Annals of the Association of American Geographers* 82(3) (1992):369–385.

Diamond, Jared. "The Environmentalist Myth: Archaeology." *Nature* 324 (1986):19–20.

———. *The Third Chimpanzee*. New York: HarperCollins, 1992.

———. "New Guineans and Their Natural World." In Stephen Kellert and Edward O. Wilson (eds.), *The Biophilia Hypothesis*. Washington, D.C.: Island Press, 1993.

Dobyns, Henry F. *Their Numbers Become Thinned: Native American Population Dynamics in Eastern North America*. Knoxville: University of Tennessee Press, 1983.

Fish, Suzanne K., Paul R. Fish, Charles Miksicek, and John Madsen. "Prehistoric Agave Cultivation in Southern Arizona." *Desert Plants* 7(2) (1985):107–112.

Flores, Mike, Fernando Valentine, and Gary Paul Nabhan. "Managing Cultural Resources in Sonoran Desert Biosphere Reserves." *Cultural Survival Quarterly* 14(4) (1990):26–30.

Gomez-Pompa, Arturo, and Andrea Kaus. "Taming the Wilderness Myth." *BioScience* 42 (1992):271–279.

Janzen, Daniel H., and Paul S. Martin. "Neotropical Anachronisms: Fruits the Gomphotheres Ate." *Science* 215 (1982):19–27.

Klett, Mark. "The Legacy of Ansel Adams." *Aperture* 120 (1990):72–73.

Martin, Paul S. "Refuting Late Pleistocene Extinction Models." In D. K. Eliot (ed.), *Dynamics of Extinction*. New York: Wiley, 1986.

Mathiot, Madeleine. *A Dictionary of Papago Usage*. Language Science Monograph 8(1). Bloomington: Indiana University Publications, 1973.

Midgley, Mary. *Beast and Man*. Ithaca: Cornell University Press, 1978.

Minnis, Paul S. "Economic and Organizational Responses to Food Stress by Non-stratified Societies: A Prehistoric Example." Ph.D. dissertation, University of Michigan, Ann Arbor, 1981.

Nabhan, Gary Paul. *Enduring Seeds*. San Francisco: North Point, 1989.

Nabhan, Gary Paul, and Sara St. Antoine. "The Loss of Floral and Faunal Story: The Extinction of Experience." In Stephen R. Kellert and Edward O. Wilson (eds.), *The Biophilia Hypothesis.* Washington, D.C.: Island Press, 1993.

National Park Service. "Revised Code of Federal Regulation for the National Park Service." *Federal Register* 52(14) (1987):2457–2458.

Oelschlaeger, Max. *The Idea of Wilderness.* New Haven: Yale University Press, 1991.

Parker, Eugene. "Forest Islands and Kayapo Resource Management in Amazonia: A Reappraisal of the *Apete*." *American Anthropologist* 94 (1992):406–427.

Rea, Amadeo R. "Resource Utilization and Food Taboos of Sonoran Desert Peoples." *Journal of Ethnobiology* 2 (1981): 69–83.

Redford, Kent H. "The Ecologically Noble Savage." *Orion* 9(3) (1985):24–29.

Redford, Kent H., and John G. Robinson. "The Game of Choice: Patterns of Indian and Colonist Hunting in the Neotropics." *American Anthropologist* 89(3) (1987):650–667.

Silko, Leslie M. "Landscape, History, and the Pueblo Imagination." In Daniel Halpern, ed., *On Nature.* San Francisco: North Point, 1987.

Soulé, Michael E. "Conservation: Tactics for a Constant Crisis." *Science* 253 (1991):744–750.

CHAPTER SEVEN

CONCEPTS OF NATURE
EAST AND WEST

STEPHEN R. KELLERT

The notion of nature as an "invention" of the human mind represents a compelling challenge to our understanding of people as social versus biological creatures. On the one hand, modern science, as well as ancient wisdom, recognizes the kinship which binds all life by a common molecular structure and genetic code. There is evidence for a shared understanding of nature in all humans, a consequence not just of the demands of physical survival, but also of emotional, intellectual, and evaluative structures associated with our dependence on the natural world (Kellert and Wilson 1993; Barkow et al. 1992). These underlying patterns intimate the existence of a common understanding of nature characteristic of all people independent of tradition, society, and geography.

On the other hand, culture and history have provided ample evidence to suggest the remarkable plasticity and relativity of human values, particularly varying conceptions of what are or should be appropriate feelings, beliefs, and behavior about the natural world. As Lynn White remarked a quarter of a century ago (1967:1205): "What people do about their ecology depends on what they think about themselves in relation to things around them. Human ecology is deeply conditioned by beliefs about our nature and destiny."

The contrast between Western (largely Judeo-Christian) and

Eastern (mainly Buddhist-Hindu) conceptions of nature constitutes an especially profound test of the universal versus relative view of human values of the natural world. While fundamental distinctions are indeed described in this chapter, essentially it will be argued that there is a biological basis for all human values of the natural world—a fundamental foundation upon which culture exercises a restricted latitude in molding human perceptions of nature. In other words, although the content and intensity of human conceptions of the natural world vary greatly in response to the formative pressures of history and culture, the underlying valuational structures remain fundamentally the same.

This chapter proceeds in three sections. The first, drawing largely on the scholarly literature, describes broad conceptions of traditional Western and Eastern perspectives of nature. This description is followed by a brief review of empirical studies of contemporary Japanese and American attitudes toward nature and wildlife. We then return to the initial question of the universal (that is, biological) versus relative (that is, cultural) basis for human values of nature. I hope to show that only a discrete set of biologically based human perspectives of nature exist, molded and shaped by history and culture, such as described in Eastern and Western societies. The deconstructionist position of nature—as solely a human creation based on power relationships—confuses content with underlying structure and thus ignores the formative influence of biologically based human valuational dependencies on the natural world.

Eastern vs. Western Conceptions

Broad descriptions of traditional Western (Judeo-Christian) and Eastern (Buddhist-Hindu) conceptions of nature can be found in the writings of various historians, philosophers, and cultural anthropologists (for example: Anesaki 1932; Callicott 1989; Callicott and Ames 1989; Higuchi 1979; Minami 1970; Murota 1986; Rolston 1989; Saito 1983; Suzuki 1973; Thomas 1983; Watanabe 1974; White 1967). A seminal essay on Western perspectives of nature was published by the historian Lynn White in *Science* in 1967. White describes traditional Judeo-Christian views of nature which collectively, he claims, are the most anthropocentric attitudes toward the natural world ever conceived, including pronounced tendencies toward human superiority, hegemony, and callousness toward nature.

A major tenet of the Judeo-Christian perspective, according to White, is the idea of a single God who created all of physical nature for human use. Moreover, humans are uniquely conceived in God's image and uniquely possess, among all creation, the capacity to achieve spiritual transcendence of their natural and physical states. The Judeo-Christian perspective, White suggests, strips nature of its sacred status, its capacity for rational consciousness, and its personhood (all fundamental principles of what he calls "pagan animism"). This desanctification of nature has permitted humans to exploit and dominate the natural world in "a mood of indifference to the [presumed] feelings of natural objects" (White 1967:1205). White further hypothesizes that Judeo-Christian beliefs might not have resulted in the widespread exploitation and destruction of nature if not for the accidental marriage of technology and science spurred by the emergence of democracy in Western society—a freedom of action, he suggests, the Western world is still struggling to cope with in its commitment to the democratic process.

The philosopher Baird Callicott (1989) further argues that the Western perspective of nature owes much to the formative influence of Greco-Roman notions. Callicott particularly emphasizes an atomistic view of nature that tends to regard the natural world as reducible to physical parts which are ultimately subject to quantitative description and mathematical laws. This perspective has fostered the Western inclination to view nature as little more than a physical fact. This viewpoint, according to Keith Thomas, has encouraged the Western scientific objective of subduing and controlling nature (1983:237): "The purpose of science was to restore to man dominion over creation. . . . Civilization and science were synonymous with the conquest of nature. . . . The whole purpose of science was that [nature] may be mastered, managed, and used in the services of human life." Thomas further argues that Western society confirms its superiority over the rest of creation by the presumption of unique human capacities for reason, speech, and moral choice. The subjugation and exploitation of nature can then be pursued, as a consequence, in an unrestricted and guiltless manner, as nature possesses no right other than that of serving some human purpose.

Traditional Eastern views are frequently depicted in nearly opposite terms. As opposed to the Western dualism unalterably distinguishing humans from nature, the Eastern perspective has been portrayed as asserting a fundamental oneness binding all creation. As the historian Roderick Nash (1989:107) suggests, the Eastern belief holds that "all beings and things, animate and inanimate, [are] permeated with divine power or spirit." Or as one Zen aphorism suggests: "All beings, even the grasses, are in the process of enlightenment." Buddhism and Hinduism are characterized as affirming a human obligation to coexist with rather than conquer nature; of promoting the belief that all living organisms share in a fundamentally analogous field of consciousness; of emphasizing principles of human compassion, respect, and reverence toward life (Kabilsingh 1987). All creation, people included, is viewed as striving toward unity, harmony, and balance, through repeated cycles of existence, ultimately tending toward a state of grace and tranquility (Tu 1989).

The Japanese historian Watanabe has described, in a 1974 *Science* paper nearly the counterpoint of Lynn White's earlier essay, a particularly harmonious and respectful Eastern conception of nature. Focusing on traditional Japanese thought, he depicts an intrinsic "love of nature resulting in a refined appreciation of the beauty of nature . . . [and a view] of man . . . considered a part of nature, and the art of living in harmony with nature the wisdom of life" (1974:280). The Japanese philosopher Murota (1986:105) further contrasts the presumed Eastern reverence for nature with the environmentally antagonistic assumptions of Western thought: "The Japanese view of nature is quite different from that of Westerners. . . . The Japanese nature is an all-pervasive force. . . . Nature is at once a blessing and friend to the Japanese people. . . . People in Western cultures, on the other hand, view nature as an object and, often, as an entity set in opposition to humankind." Mariani (1971:19) similarly asserts "a Japanese [Eastern] love and reverence for nature . . . , a relationship . . . based on feelings of awe and respect . . . in harmony with an understanding of nature's totality."

These characterizations of Eastern and Western conceptions of nature may be thought of as idealized depictions partially indicative of the behavior and beliefs of particular individuals and groups. How much these broadly construed values actually explain human thought and action, however, is unclear. Moreover, these Eastern and Western conceptions of nature are associated with traditional religion and culture, and their relevance to the beliefs and behavior of people in modern society, particularly in highly industrialized societies, leaves room for doubt. Indeed, in contrast to the foregoing descriptions of highly positive Eastern attitudes toward nature, modern Japan and China have been cited for their poor conservation record—including widespread temperate and tropical deforestation, excessive exploitation of wildlife products, indiscriminate and damaging fishing practices, and widespread pollution (Kamei 1983; Oyadomori 1985; Saiki 1988; Sneider 1989; Sun 1989; Taylor 1990; Upham 1979).

Studies of Modern Attitudes

This paradox of a presumably Eastern respect for nature against a record of contemporary environmental destruction prompted the author to conduct a national study of attitudes toward nature and wildlife in Japan. A national investigation had already been conducted in the United States. Both studies employed similar concepts and methodology, although the year of the study and the data collection strategies differed (Kellert 1980, 1985, 1989, 1991, 1993).

The Method

The principal methodological procedure was a general population survey. Nearly two hundred questions were included in the two surveys covering attitudes, knowledge, and behavior toward nature, mainly wildlife. Although the questions in both surveys were largely identical, the wording was sometimes changed and different but taxonomically parallel species were sometimes employed to accommodate for cultural and biogeographical variations between the two countries. In both studies, personal interviews were conducted with randomly selected members of the adult population (eighteen years of age and older). The American study involved 3,107 interviews in the forty-eight contiguous states and Alaska. The Japanese study, which consisted of 450 interviews with respondents residing in Tokyo and three widely distributed rural locations, employed stratified random sampling to obtain sampling quotas based on age, gender, and education using census statistics. The Japanese study also included in-depth, open-ended interviews with fifty informants chosen

Table 7.1
BASIC ATTITUDES TOWARD ANIMALS

Term	Definition
Naturalistic	Primary interest and affection for wildlife and the outdoors
Ecologistic	Primary concern for the environment as a system and the relationships between wildlife species and natural habitats
Humanistic	Primary interest and strong affection for individual animals such as pets or large wild animals with strong anthropomorphic associations
Moralistic	Primary concern for the right and wrong treatment of animals and strong opposition to overexploitation and cruelty toward animals
Scientistic	Primary interest in the physical attributes and biological functioning of animals
Aesthetic	Primary interest in the physical and symbolic appeal of animals
Utilitarian	Primary interest in the practical value of animals or in the subordination of animals for the practical benefit of people
Dominionistic	Primary interest in the mastery and control of animals
Negativistic	Primary orientation an active avoidance of animals due to dislike or fear
Neutralistic	Primary orientation a passive avoidance of animals due to indifference or lack of interest

from a large number of persons recommended for their extensive knowledge of Japanese attitudes toward nature.

A typology of ten basic attitudes toward nature and wildlife constituted the primary conceptual tool. This typology is thought to reflect universal and fundamental values of the natural world. One-sentence definitions of these basic attitudes are indicated in Table 7.1. Attitude scales were constructed, based on cluster and factor analysis results, involving four to nine questions indicative of the underlying attitude. As this methodology did not yield an adequate aesthetic scale in either country or independent scales of the neutralistic and negativistic attitudes, the two attitudes were combined into one overall scale. Twenty-five true/false questions were used to construct a knowledge of animals scale.

The similarity of methods and concepts employed in the two studies provided a reasonable basis for comparing the results. Important methodological differences and varying data collection strategies, however, suggest caution in interpreting the findings. Before reviewing the results, we should note certain socioeconomic and biogeographical characteristics of modern Japan and the United States.

The United States encompasses a much larger territory than Japan. The American population is also twice as large, with approximately 250 million people compared to Japan's roughly 125 million. Because of its size, the United States has a population density of 25 people per square kilometer compared to 311 in Japan (Yano 1984). In relation to arable land, the Japanese population is 2,256 per square kilometer compared to 103 in the United States (Marsh 1987). More than three-quarters of the Japanese population resides in urban areas, resulting in less than 50 people per square kilometer in the roughly two-thirds of the country covered by mountains.

Despite Japan's high population density, the country has a surprisingly large amount of biological diversity, as measured by numbers of species. Japan has more than twice the number of mammalian, avian, and invertebrate species as does the comparably sized Great Britain and Ireland combined (Japan Environment Agency 1982). This degree of biological diversity, as well as the unusually large number of endemic species, is due largely to Japan's relative geographic isolation and variation associated with its highly mountainous terrain, the country's broad latitudinal variation (from the boreal forests of Hokkaido to the neotropics of Okinawa), and its large number of islands (with a cumulative coastline of nearly 33,000 kilometers, possibly the second-longest shoreline in the world according to Trewartha 1965).

The 1992 GNP of the United States was approximately $6 trillion, nearly twice the size of Japan's (CIA 1992). Japan has experienced very rapid economic growth, however, as indicated by its 1960 GNP of only $300 billion. Despite this largely industrial expansion, Japan has a higher proportion of its population engaged in agriculture (9 percent) than does the United States (3 percent).

The Results
The relative rank order and basic attitude scores of respondents in the United States and Japan are indicated in Table 7.2. Both countries had similarly high humanistic scores, suggesting the importance of strong emotional attachments to individual animals, single

Table 7.2
RANK ORDER AND STANDARDIZED MEAN SCORES OF ATTITUDE TOWARD NATURE
IN THE UNITED STATES AND JAPAN

Rank	United States	Japan
1	Humanistic (0.38)	Humanistic (0.37)
2	Moralistic (0.275)	Negativistic (0.31)
3	Negativistic (0.26)	Dominionistic (0.28)
4	Utilitarian (0.23)	Naturalistic (0.22)
5	Ecologistic (0.215)	Utilitarian (0.22)
6	Naturalistic (0.20)	Moralistic (0.18)
7	Dominionistic (0.13)	Ecologistic (0.16)
Knowledge scale:	53	48

Note: United States data collected in 1977 ($n = 3,107$); Japan data collected in 1987 ($n = 450$). Differences in dominionistic, ecologistic, moralistic, negativistic, and knowledge scores between the United States and Japan are statistically significant.

species, and specific elements of the landscape. This humanistic perspective was typically directed at environmental features distinguished by their aesthetic appeal, cultural significance, and historic familiarity. Moderately pronounced naturalistic scores in both countries suggest an interest in directly experiencing these preferred aspects of the natural environment. Relatively high utilitarian scores in both countries indicate a strong pragmatic orientation to nature. Comparatively high negativistic scores, especially in Japan, intimate an attitude of indifference toward aspects of the natural environment lacking valued emotional, cultural, utilitarian, or historic appeal.

But a number of significant attitudinal variations were also observed. Significant dominionistic differences suggest a Japanese public far more inclined than the American to emphasize control over nature, particularly when the environmental features possess unusual aesthetic and emotional appeal. Highly significant ecologistic and moralistic scale differences reveal an American public far more likely than the Japanese to express concern about the ethical treatment of animals, support for environmental protection, and interest in ecological functioning. These ecologistic and moralistic differences are further reflected in age and educational group results. In both countries, better educated and younger persons reveal much greater appreciation of and interest in nature and wildlife than those of less ed-

Table 7.3

Analysis of Variance and Multiple Classification Analysis for Attitude and Knowledge Scales by Age in the United States ($N = 3{,}107$)

Attitude	Deviation from Grand Mean After Adjusting for Independent and Covariate Variables				
	18–25	26–35	36–55	56–75	76+
Naturalistic (sig. of $F = 0.000$)	0.52	0.36	−0.09	−0.54	−0.49
Ecologistic (sig. of $F = 0.039$)	0.07	0.10	0.12	−0.29	−0.08
Humanistic (sig. of $F = 0.000$)	0.71	0.16	−0.15	−0.37	−0.67
Moralistic (sig. of $F = 0.010$)	0.44	0.49	−0.32	−0.30	−0.53
Scientistic (sig. of $F = 0.000$)	0.30	0.28	0.10	−0.30	−0.38
Utilitarian (sig. of $F = 0.000$)	−1.14	−0.88	0.26	1.15	1.43
Dominionistic (sig. of $F = 0.411$)	−0.08	−0.05	0.11	−0.08	0.22
Negativistic (sig. of $F = 0.000$)	−0.69	−0.45	0.25	0.50	0.74
Knowledge (sig. of $F = 0.000$)	−2.30	−0.49	0.84	1.57	−3.12

Note: Multiple classification analysis is a statistical technique based on analysis of variance which allows one to determine how much certain categories of a variable are different from the overall population mean adjusted for the effects of other variables. In other words, this technique allows us to determine which categories of a variable contribute the most to the overall significance of the variable—for example, which age or educational groups are most related to a particular scale after other variables have been taken into account.

ucation and the elderly, as suggested by the results of Tables 7.3 through 7.6. On the other hand, the differences in moralistic and ecologistic scores among Japanese age and education groups are insignificant—in striking contrast to those found in the United States as well as in Germany (Kellert 1993). College-educated and younger Americans were far more likely to express ethical and ecological concerns than were their counterparts in Japan. It may be relevant to note that less than 1 percent of the Japanese sample reported membership

Table 7.4
ATTITUDES OF VARYING AGE GROUPS IN JAPAN

Attitude	18–35	36–44	45–64	65+
Naturalistic ($p = 0.01$)	0.225	0.28	0.22	0.20
Ecologistic/Scientistic ($p = 0.11$)	0.17	0.21	0.17	0.155
Negativistic ($p = 0.0006$)	0.28	0.24	0.31	0.365
Humanistic ($p = 0.07$)	0.40	0.40	0.32	0.34
Dominionistic ($p = 0.01$)	0.24	0.24	0.26	0.30
Utilitarian ($p = 0.09$)	0.21	0.20	0.23	0.24
Moralistic ($p = 0.40$)	0.18	0.22	0.20	0.19
Knowledge ($p = 0.03$)	0.495	0.498	0.452	0.463

in a conservation organization compared to 11 percent in the United States. In a related finding, the United Nations Environmental Program (Japan Prime Minister's Office 1989:7) found that of fourteen surveyed countries Japan rated the "lowest in environmental concern and awareness."

A less factual understanding of nature and animals among the Japanese is reflected in significantly lower knowledge scores than in the United States. Based on the knowledge question results, the American public appears to possess a greater understanding of basic biological processes, while the Japanese public has a better knowledge of economically valuable species (Table 7.7).

The Japanese statistical results were complemented in many important ways by the qualitative findings derived from the in-depth interviews with fifty key informants. Most of the respondents indicated the Japanese tend to place greatest emphasis on the experience and enjoyment of nature in highly structured circumstances. The objective, as one informant suggested, was to capture the presumed essence of a natural object by adhering to strict rules of seeing and ex-

Table 7.5
ANALYSIS OF VARIANCE AND MULTIPLE CLASSIFICATION ANALYSIS FOR ATTITUDE
AND KNOWLEDGE SCALES BY EDUCATION IN THE UNITED STATES ($N = 3,107$)

Attitude	Deviation from Grand Mean After Adjusting for Independent and Covariate Variables				
	<8th Grade	9–11	12th and Voc/Tech	College	Graduate
Naturalistic (sig. of $F = 0.000$)	−0.61	−0.54	−0.20	0.36	1.13
Ecologistic (sig. of $F = 0.000$)	−0.35	−0.58	−0.21	0.24	1.38
Humanistic (sig. of $F = 0.728$)	−0.27	−0.04	0.02	0.08	−0.01
Moralistic (sig. of $F = 0.000$)	−0.27	−0.49	−0.33	0.36	1.32
Scientistic (sig. of $F = 0.000$)	−0.21	−0.22	−0.19	0.17	0.83
Utilitarian (sig. of $F = 0.000$)	0.87	0.31	0.12	−0.29	−0.85
Dominionistic (sig. of $F = 0.019$)	0.20	0.14	0.11	−0.13	−0.45
Negativistic (sig. of $F = 0.000$)	0.95	0.33	0.18	−0.38	−0.99
Knowledge (sig. of $F = 0.000$)	−5.10	−3.36	−1.27	2.36	7.73

periencing intended to best express the centrally valued feature. Rarely did this admiration extend beyond single species or particular landscapes to a broad appreciation of the natural world or the ecological processes associated with it. Environmental features falling outside the especially valued aesthetic and symbolic boundaries of preferred natural objects tended to be ignored, judged irrelevant, or perceived as unappealing.

This restricted appreciation of nature was described by many informants as largely emotional with little ecological or biological basis. One informant referred to it as a Japanese love of "seminature," somewhat domesticated and tamed. Another respondent emphasized a Japanese preference for the contrived, highly abstract, and sym-

Table 7.6
ATTITUDES OF VARYING EDUCATION GROUPS IN JAPAN

Attitude	Middle School	High School	Voc/Tech Jr. College	College
Naturalistic ($p = 0.0001$)	0.19	0.22	0.25	0.32
Ecologistic/Scientistic ($p = 0.27$)	0.16	0.17	0.21	0.19
Negativistic ($p = 0.0001$)	0.385	0.29	0.32	0.17
Humanistic ($p = 0.75$)	0.35	0.37	0.40	0.38
Dominionistic ($p = 0.23$)	0.28	0.25	0.24	0.23
Utilitarian ($p = 0.001$)	0.27	0.22	0.19	0.18
Moralistic ($p = 0.51$)	0.18	0.19	0.21	0.21
Knowledge ($p = 0.001$)	0.41	0.479	0.512	0.546

Table 7.7
SELECTED KNOWLEDGE QUESTIONS (% CORRECT)

Question	Japan	U.S.
Most insects have backbones.	45	57
Snakes have a layer of "slime" to move more easily.	34	52
All adult birds have feathers.	49	63
Spiders have ten legs.	37	50
A seahorse is a kind of fish.	40	71
Salmon breed in fresh water but spend most of their life in salt water.	76	66

All the following are mammals:
    Japan: impala, tanuki, iguana, killer whale
    U.S.: impala, muskrat, iguana, killer whale
	20	40

bolic rather than the realistic experience of nature or wildlife. Many respondents suggested a Japanese motivation to touch nature from a safe distance or, as another informant remarked, to isolate favored environmental features and "freeze and put walls around them." One respondent described this tendency as a Japanese inclination "to go to the edge of the forest, to view nature from a mountaintop, but not to enter or immerse oneself in wildness or the ecological understanding of natural settings."

A Japanese affinity for nature was identified, therefore, but typically as a restricted idealistic recreation or artistic rendering of especially valued attributes. This refined appreciation of nature involved considerable abstraction and a concept of perfection derived from prescribed rules and assumptions of harmony, order, and balance in the natural world. One informant characterized it as "a tradition of stealing aspects of nature and creating an art form around it." Donald Ritchie (1971:13) similarly suggests: "The Japanese attitude toward nature is essentially possessive. . . . Nature is not natural . . . until the hand of man . . . has properly shaped it." Grapard (1985:243) observes: "What has been termed the Japanese love of nature is actually the Japanese love of cultural transformations . . . of a world which, if left alone, simply decays." And Saito concludes that this perspective is largely escapist and unlikely to foster much of a conservation ethic (1983:192): "Nature is not lived or respected for its own sake but because it allows one to escape. . . . This appreciation of nature not only implies an anthropocentric attitude . . . but also suggests [an] ineffectiveness in generating an ethically desirable justification for protecting nature."

The Meaning
These research findings do not confirm the idealized descriptions reported earlier of traditional Eastern and Western conceptions of nature. Although considerable affection and appreciation for nature are expressed in the United States and Japan, these perspectives are typically narrow and directed at only a small segment of the natural world. Moreover, this admiration lacks, especially in Japan, an ethical or ecological perspective or strong support for nature conservation and protection.

A number of explanations could be offered for this difference between the traditional depictions of Eastern and Western conceptions of nature and the findings reported here. It might be suggested that traditional Eastern perspectives of nature have largely been replaced in modern Japan by Western views. Additionally, it could be asserted that both modern Japan and the United States share a conception of nature dominated by the values of modern industrial, technological, and scientifically oriented society.

The possibility that modern Japan has shifted from a traditional Eastern/Buddhist conception of nature to a predominantly Western one has been suggested by various scholars. Watanabe (1974:281), for example, cited earlier for his view of a "traditional Japanese way of looking at nature [involving] an affinity and sympathy with all living things," goes on to assert that "these . . . sentiments have been rapidly fading in Japan since the hasty introduction of modern science and technology." Most observers of modern Japan generally cite the Meiji period, some two centuries ago, as the beginning of the country's westernization (Bunge 1981; Higuchi 1979). Callicott and Ames (1989:280), acknowledging the apparent contradiction between Asia's current environmental destruction and the notion of a traditional Eastern reverence for nature, also lay the blame for this change on the ascendancy of Western views of the natural world: "Contemporary environmental misdeeds perpetrated by Asian peoples today can in large measure be attributed to the intellectual colonization of the East by the West. . . . All . . . Asian environmental ills . . . are either directly caused by Western technology . . . or aggravated by it. . . . Modern technology is embedded . . . in the dominant Western paradigm."

These scholars therefore assert that a Western-oriented, and perhaps more inclusive, industrial/technocratic value system has emerged in modern Asia at variance with its traditional Buddhist attitudes of respect for nature. This explanation, while appealing in many ways, strikes one as incomplete and simplistic. Indeed, the views of contemporary Japan revealed in the research reported here are quite compatible with traditional Eastern conceptions of nature. These conceptions, highly abstract and idealized, typically involve little empirical understanding of nature or ecological considerations of the natural world. They rarely provide much explicit support for nature conservation, beyond a relatively indiscriminate covenant not to cause suffering and to be compassionate. Moreover, traditional Eastern attitudes toward nature often encourage passivity, even fatalism, toward a natural world depicted as all-powerful and beyond human capacity to control or grasp, let alone conserve or regulate. While the ascendance of Western values and technology might provide some understanding for the emergence of less benign attitudes toward nature in modern Asia, a more discerning examination of traditional Eastern conceptions indicates important limitations in this perspective as an adequate basis for an ecological ethic of concern for the natural world and its conservation.

Toward a Cultural Synthesis

This chapter has largely considered the human capacity, through culture and religion, to mold and shape what appear to be unique constructions of the natural world. While these social articulations intimate the possibility of environmental values being largely the creation of history and culture, these constructions are viewed as manifest only within the restricted context dictated by human biology and evolution. In other words, I believe there is a biological basis for all human values of nature. These biologically based constructs reflect a range of physical, emotional, and intellectual needs associated with human survival and evolutionary adaptation (Kellert and Wilson 1993). Societies can develop varying cultural strategies for expressing these needs, but they cannot alter the underlying structural imperative.

Eastern and Western conceptions of nature should be viewed, in other words, as indicative of the content, direction, and intensity which culture gives to diverse biologically based values of nature. This is not to suggest, however, that all perspectives of nature are equally valid or functionally adaptive. Cultural perspectives can profoundly influence the potential for humans to achieve a positive, satisfying, and healthy existence. Societies, like individuals, possess the capacity for functional expression as well as harmful distortion of basic human physical, affective, and cognitive needs. Just as individuals can act in self-enhancing and self-defeating ways, cultures can develop both healthy and maladaptive beliefs in relation to nature. The degradation of the human dependence on the natural world, whether by individuals or by societies, increases the likelihood of a deprived and diminished existence—not just materially but in a wide variety of emotional, intellectual, and evaluative respects.

Neither Eastern nor Western societies are intrinsically inferior or superior in their perspectives of nature. This chapter has endeavored to demonstrate, both conceptually and empirically, that both cultural viewpoints reflect functional and dysfunctional attitudes toward the natural world. From a positive perspective, both cultural traditions have embedded within their conceptions of nature the seeds of a powerful ethic of appreciation, respect, and concern for the conservation of nature. From the East, we derive an enhanced compassion and appreciation for life, a profound intuition of nature's oneness, and the willingness to exist in harmony and balance with the natural world. From the West, we obtain the inclination to understand nature empirically, a tradition of environmental stewardship, and the belief in wisely managing and controlling the natural world.

The urgency of achieving a more culturally positive relation to the natural world has become especially evident in modern times. We are confronted with varying twenty-first-century scenarios of environmental apocalypse, from massive impoverishment of life on earth to widespread atmospheric degradation to extensive resource depletion to a rain of toxic contaminants. The deconstructionist notion that all cultural perspectives of nature possess equal value is both biologically misguided and socially dangerous. More than ever, modern society needs to fashion a cultural synthesis drawing from the best of the great Western, Eastern, and aboriginal traditions in seeking a more benign and nurturing relationship to nature. In the process, it may be possible for us to understand how through our many-faceted relation to the natural world can emerge the capacity for a fuller humanity.

References

Anesaki, M. *Art, Life and Nature in Japan*. Rutland, Vt.: Tuttle, 1932.

Barkow, J., L. Cosmides, and J. Tooby (eds.). *The Adapted Mind: Evolutionary Psychology and the Generation of Culture*. Oxford: Oxford University Press, 1992.

Bunge, F. (ed.). *Japan: A Country Study*. Washington, D.C.: Superintendent of Documents, 1981.

Callicott, J. B. *In Defense of the Land Ethic: Essays in Environmental Philosophy.* Albany: State University of New York Press, 1989.

Callicott, J. B., and R. T. Ames (eds.). *Nature in Asian Traditions of Thought.* Albany: State University of New York Press, 1989.

Central Intelligence Agency (CIA). *Handbook of Economic Statistics.* Washington, D.C.: Superintendent of Documents, 1992.

Grapard, A. "Nature and Culture in Japan." In M. Tobias (ed.), *Deep Ecology.* San Diego: Avant Books, 1985.

Higuchi, K. *Nature and the Japanese.* Tokyo: Kodansha International, 1979.

Japan Environment Agency. *The Natural Environment of Japan.* Tokyo: Japan Environment Agency, 1982.

Japan Prime Minister's Office. *The Law and Related Standards for the Protection of Animals.* Tokyo: Prime Minister's Office, 1989.

Kabilsingh, C. "How Buddhism Can Help Protect Nature." In N. Nash (ed.), *Tree of Life: Buddhism and Protection of Nature.* Hong Kong: Buddhism Protection of Nature Project, 1987.

Kamei, N. "Clearing the Clouds of Doubt." *Mainstream* (1983):22–25.

Kellert, S. R. "Contemporary Values of Wildlife in American Society." In W. Shaw and I. Zube (eds.), *Wildlife Values.* Ft. Collins, Colo.: U.S. Forest Service, 1980.

———. "American Attitudes Toward and Knowledge of Animals." In M. Fox and L. Mickley (eds.), *Advances in Animal Welfare Science.* Washington, D.C.: Humane Society of the United States, 1985.

———. "Perceptions of Animals in American Culture." In R. Hoage (ed.), *Perceptions of Animals in American Culture.* Washington, D.C.: Smithsonian Press, 1989.

———. "Japanese Perceptions of Wildlife." *Conservation Biology* 5 (1991):297–308.

———. "Attitudes, Knowledge, and Behavior Toward Wildlife Among the Industrial Superpowers: United States, Japan, and Germany." *Journal of Social Issues* (1993a):42.

———. "The Biological Basis for Human Values of Nature: A Typology." In S. R. Kellert and E. O. Wilson (eds.), *The Biophilia Hypothesis*. Washington, D.C.: Island Press, 1993b.

Kellert, S. R., and E. O. Wilson (eds.). *The Biophilia Hypothesis*. Washington, D.C.: Island Press, 1993.

Mariani, F. *Japan: Patterns of Continuity*. New York: Harper & Row, 1971.

Marsh, J. *Marine Parks in Japan*. Ottawa: Environment Canada, Parks, 1987.

Minami, H. *Psychology of the Japanese People*. Honolulu: East-West Center, 1970.

Murota, Y. "Culture and the Environment in Japan." *Environmental Management* 9 (1986):105–112.

Nash, R. *The Rights of Nature: A History of Environmental Ethics*. Madison: University of Wisconsin Press, 1989.

Oyadomori, N. "Politics of National Parks in Japan." Ph.D. dissertation, University of Wisconsin, 1985.

Ritchie, D. *The Island Sea*. Tokyo: Weatherhill, 1971.

Rolston, H. "Respect for Life: Can Zen Buddhism Help in Forming an Environmental Ethic?" In *Zen Buddhism Today*. Annual Report of the Kyoto Zen Symposium. Kyoto: Kyoto Seminar for Religious Philosophy, 1989.

Saiki, M. "Modernization of Japan with the Insight of Conservation." Unpublished manuscript, Department of Forestry, University of Maine, 1988.

Saito, Y. "The Aesthetic Appreciation of Nature: Western and Japanese Perspectives and Their Ethical Implications." Ph.D. dissertation, University of Wisconsin, 1983.

Sneider, D. "Japan Assailed for Practices That Damage Global Environment." *Christian Science Monitor*, 11 July 1989.

Sun, M. "Japan Prodded on the Environment." *Science* 241 (1989):46.

Suzuki, D. T. *Zen and Japanese Culture.* Princeton: Princeton University Press, 1973.

Taylor, R. "On the Japanese Love of Nature." Unpublished manuscript, 1990.

Thomas, K. *Man and the Natural World.* New York: Pantheon Books, 1983.

Trewartha, G. *Japan: A Geography.* Madison: University of Wisconsin Press, 1965.

Tu, Wei-ming. "The Continuity of Being: Chinese Visions of Nature." In J. B. Callicott and R. T. Ames (eds.), *Nature in Asian Traditions of Thought.* Albany: State University of New York Press, 1989.

Upham, F. "After Minimata: Current Prospects and Problems in Japanese Environmental Litigation." *Ecology Law Quarterly* 8 (1979):213–268.

Yano, I. *Nippon: A Chartered Survey of Japan.* Tokyo: Kokuseisha Corporation, 1984.

Watanabe, H. "The Conception of Nature in Japanese Culture." *Science* 183 (1974):279–282.

White, L. "The Historic Roots of Our Ecologic Crisis." *Science* 183 (1967):1203–1207.

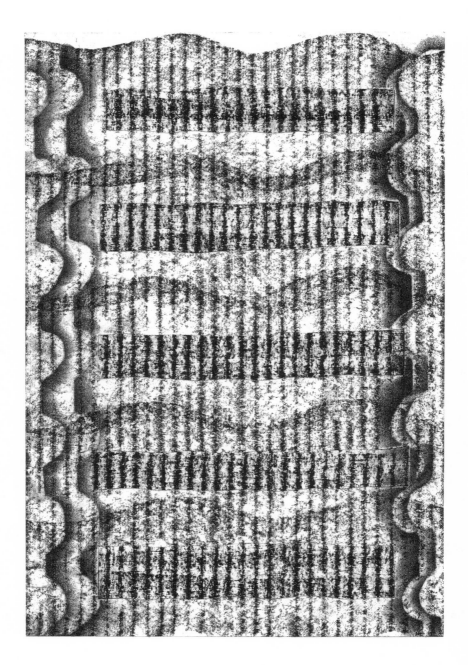

CHAPTER EIGHT

RESOLUTE BIOCENTRISM: THE DILEMMA OF WILDERNESS IN NATIONAL PARKS

DAVID M. GRABER

The organizing myth of America's national parks today is wilderness. It was forged in the debates between John Muir and Gifford Pinchot as the nineteenth century wound down. This myth, however, had little to do with the actual establishment of the first national parks: Yellowstone, Yosemite, and Sequoia. These parks were meant to protect spectacles of nature—geysers, waterfalls, and huge trees— and promote them as the virtuous attributes of a young nation lacking the constructed marvels of the Old World. This mandate for "object protection" is enshrined in the National Park Service Act of 1916: "to conserve the scenery and the natural and historic objects and the wild life therein and to provide for the enjoyment of the same in such manner and by such means as will leave them unimpaired for the enjoyment of future generations." In fact, the wilderness myth so clearly articulated by Muir simply gestated as policy for the better part of a century before its emergence as the basis upon which national parks (the large, mostly western, nature parks) are now managed.

The unifying principle of national park management today is the perpetuation of native ecosystem elements and processes. That is: keep all the native species; seek the free play of fire, water, wind, predation, and decomposition, the processes of the ecosystem; fend off alien organisms; and then permit the ecosystem to sort itself out. As management policy, it is rarely if ever fully expressed, but it has been a goal at which managers could aim (Graber 1983).

The passage of the first Wilderness Act in 1964 formalized and enshrined the notion of wilderness into law. In subsequent years, the bulk of the lands in the large western "natural" national parks, much of the unroaded portions of national forests and Bureau of Land Management lands, as well as some eastern parklands, became legal wildernesses. The law prevents nearly all mechanized use and development of wilderness, but in most cases it permits grazing, hunting, and fishing where these activities were already permitted. Over the years the practice of "wilderness management" has evolved in the federal agencies, occupying itself with such issues as social carrying capacities, waste-handling techniques, trail design, wilderness fire management, minimum-impact camping, acceptable practices in rock climbing, and the appropriate use of helicopters over and in wilderness.

Wilderness has taken on connotations, and mythology, that specifically reflect latter-twentieth-century values of a distinctive Anglo-American bent. It now functions to provide solitude and counterpoint to technological society in a landscape that is *managed* to reveal as few traces of the passage of other humans as possible. Contemporary wilderness visitors are just that. Unlike the hunters and gatherers who preceded them on the land, moderns who enter wilderness do so not to live on the land, nor to use it, but rather to experience it spiritually. The ecosystem is defined on its own terms, but this wilderness is a social construct.

Protecting the spiritual values of wilderness for its users has been the principal aim of wilderness managers. These are largely "human-on-human" effects: social crowding, conflicts between hikers and equestrians, litter, feces, campfire rings. They affect how visitors experience wilderness but have a minimal effect on wild ecosystem functions. For the most part, neither managers nor visitors are aware of the ways in which the wilderness landscape has been altered by former aboriginal activities (hunting, clearing, burning, agriculture), past and present local landscape alterations (game extinctions, logging, alien introductions, riparian destruction by grazing), or systemwide human impacts (habitat fragmentation, air pollution, suppression of native fire regimes, climate change).

Fire Management in Sequoia: The Indian Dilemma

The evolution of fire management in the national parks recapitulates in many ways the evolution of the park wilderness principle. Ecologists recognized early in the century that the establishment and maintenance of certain plant species and vegetation types depend on periodic fire. Among national parks, this is particularly obvious in the Everglades in the Southeast, where the pine/sawgrass community requires frequent fire, and in the Sierra Nevada—Sequoia, Kings Canyon, Yosemite—vegetation from foothill chaparral to subalpine forests depends on fire as well. Nonetheless, the larger social assumption that fire, because it destroys life and property, should be suppressed at all costs determined how it would be managed in national forests and parks until the 1960s.

At that time, scientists and then park managers began to realize that in the mixed-conifer forests of the Sierra Nevada, fire suppression had prevented the reproduction of the giant sequoia (*Sequoiadendron giganteum*) and gradually led to the dense ingrowth of shade-tolerant species such as white fir (*Abies concolor*) that ultimately would transform the naturally frequent and thus relatively cool Sierran fires into fearsome crown fires that could destroy the ancient monarch sequoias themselves. The practice of intentional burning to reduce fuel accumulations—"prescribed management fire"—was born. Ring scars in old giant sequoias (Kilgore and Taylor 1979) suggested that, in the most recent millennium at least, the high frequency of fires could not be explained by the contemporary ignition rate from lightning. They suggested that the extra fire had been produced by aboriginal burning.

During the 1970s and 1980s, park managers and scientists struggled to define fire management—a combination of prescribed management fire, "prescribed natural fire" (lightning-caused fires allowed to burn under certain constraints), and the suppression of all other ignitions, natural or human in origin, that failed to occur "under prescription" (Graber 1985). There were several difficulties, however. In the beginning (Leopold et al. 1963), the purpose of this program was to restore a more open forest structure and stimulate reproduction of the giant sequoia, both of which were presumed to have suffered severely during a century of fire suppression, while at the same time reducing dangerous levels of fuels (living and dead wood) that had accumulated during that same period. These goals, largely concerned with forest structure and aesthetics, may reflect an instinctive human preference for "open and parklike" forests over "dog-hair thickets" of white fir (Leopold et al. 1963), as well as a preference for the heroic giant sequoia—after which the park had been named—over the "pissfir" despised by foresters for its inferior wood. The work of Kilgore and Taylor (1979), Lewis (1973), and others suggested that this "desired" forest state would require supplemental ignitions to substitute for the centuries of aboriginal forest burning.

An early challenge to straightforward prescribed burning was presented by Bonnicksen and Stone (1982), who argued that burning a forest possessing a greatly altered structure on account of a century of fire suppression would lead to new artifices of both structure and fire regime. The basis of their objection was that fire should be viewed as a tool to create a structural result—in their case, the forest that would have been present had not a century of fire suppression intervened. A second objection was introduced by Eric Barnes (National Park Service files), who objected to the aesthetic changes in giant sequoia stands, especially blackened bark, produced by prescribed fire. His position was based on visitors' perceptions of the park's "prime resource," the monarch bigtrees.

More fundamental than these objections, however, were conceptual weaknesses in the dual objectives of prescribed fire: fuel reduction (that is, safety) and improved forest structure. These objections became apparent as the technology of prescribed burning improved and the paleoecology of the Sierra Nevada became better understood (Anderson 1990; Graumlich 1993; Swetnam 1993). Mimicking aboriginal/lightning fire patterns presented two apparent dilemmas: Indians had been present in the Sierra for less than twenty thousand years, a far shorter period than the present array of species; moreover, as cultural creatures, the Sierran Indians and their landscape-altering practices no doubt were in a flux of substantially greater rapidity than ecological time. Had they remained undisturbed by Europeans, it is unlikely they would have continued pursuing deer, collecting acorns, and lighting fires for millennia to come.

The second problem was climate change. Revealed indirectly through tree rings (Graumlich 1993) and pollen cores (Anderson 1990), it had been expressed at the millennial scale as dramatic variations in the dominance of species and even physiognomic groups at the present montane sites of mixed-conifer forest and at the century scale as equally dramatic differences in fire intensities and frequencies. Given these dynamic forces, what was the objective of administrative burning? The logic of attempting to simulate a fire regime produced by a dynamic aboriginal culture operating in a dynamic climatic regime began to fade by the late 1980s (Parsons et al. 1986).

Nonetheless, some sort of fire management decision was required: the contemporary park forests were changing in ways reflecting the intrusions of industrial culture. Fire-dependent species were declining while the forests as a whole were becoming more flammable and dangerous, and less attractive. The course selected was to use prescribed management fire as a corrective measure only—to reverse the fuel accumulation of the past century—and then to permit natural forces—lightning and climate—to determine fire regime and forest structure henceforward. Thus the hard decisions would be left to nature and the National Park Service would be managing for wildness, not some "desired state." Yet to be resolved are developing conflicts between smoke-producing Sierra fires and regional air quality standards. There is the growing suspicion, too, that the important events in the formation of Sierran vegetation architecture naturally occur not during frequent low-intensity fires but when the century-scale combinations of weather and vegetation produce large, hot, and dangerous fires that neither fire managers nor society are ready to accept. Most worrisome of all, the parks' fire management program has yet to reestablish a fire frequency even 10 percent of the rate recorded in tree rings.

The Biodiversity Problem

Fundamental to the notion of a wilderness park is the idea that it must possess its aboriginal array of native biota and associated ecosystem processes. American national parks have, in the latter part of the twentieth century, come to serve as its preeminent nature reserves; most international biosphere reserves in this country are also national parks. The nature these reserves are intended to protect is the indigenous biological diversity of the region. Yet increasingly, national park ecosystems are being transformed by the introduction of alien plants and animals and, secondarily, by the extinction of native populations as a result of these alien introductions. As the following examples illustrate, not only is local biodiversity being swamped but the management keys to retarding its loss conflict with the park notion of wild nature.

In Shenandoah National Park in the Virginia Appalachian Mountains, pests and pathogens from Eurasia have devastated the populations of American chestnut (*Castanea dentata*), American elm (*Ulmus americana*), eastern hemlock (*Tsuga canadensis*), and eastern dogwood (*Cornus florida*), while the gypsy moth (*Lymantria dispar*) continues to weaken and kill a variety of oaks and other hardwood trees. Stresses induced by ozone and acidic pollutants may be potentiating these epibiotics. This loss of a substantial portion of the dominant native flora has reduced the populations of birds and mammals that depend on the fruits and seeds and the cover these trees once produced. And of course the impacts ramify through insect populations, those that depend on them, and so on. Parallel but less severe losses are taking place in Great Smoky Mountains National Park in the southern Appalachians. Although these ecological changes are acknowledged by park management to be a loss of much of the wild character of the areas, no solutions present themselves.

The Sierra national parks, Yosemite, Kings Canyon, and Sequoia, have suffered comparatively little human-induced species turnover (Macdonald et al. 1988). One notable exception is the foothills below about forty-five hundred feet, where—as for most of the rest of California's low-elevation wildlands—Eurasian annual grasses and some dicots have virtually replaced the native herbaceous species. The weedy annuals (such as wild oat, *Avena fatua*; cheatgrass, *Bromus tectorum*; filaree, *Erodium cicutarium*) were largely introduced and established during the mid-nineteenth-century period of intensive cattle and sheep grazing, including the present parks, that overwhelmed native herbs unadapted to intensive grazing. There is reason to suspect that these introduced annuals have changed fire frequencies, interfere with recruitment of the native woody plants with which they occur, and have a significant impact on the birds, small mammals, and reptiles indigenous to the foothills. There are no practical means yet known to permanently remove these aggressive weedy aliens over large areas. Nor is there good information on what the native herbaceous layer consisted of, should an opportunity arise to restore it.

A similar situation obtains in Channel Islands National Park, where long and continuing grazing has destroyed the native vegetation to the point where many endemic plant species have been reduced to small numbers of individuals. Elimination of grazing might be sufficient to permit reestablishment of some healthy native vegetation, but not without the control of the aliens. Because they are islands, however, the Channel Islands may lend themselves more readily to attempts at extirpating the weedy annuals.

The large Hawaiian national parks, Hawaii Volcanoes and Haleakala, have experienced substantial replacement of the native flora and fauna by introduced species, a process that continues at a rapid pace. Avian malaria carried by accidentally introduced mosquitoes has eliminated most of the native avifauna except in the mountains, while dozens of alien bird species have been introduced intentionally or have escaped from captivity. Alien herbs, shrubs, trees, and vines (christmasberry, *Schinus terebrinthifolius*; strawberry guava, *Psidium cattleianum*; blackberry, *Rubus argutus*; banana poka, *Passiflora mollisima*) now constitute a significant fraction of the parks' biomass, leading to continuing efforts to find pests and diseases that can be introduced to control them (Smith 1989). Introduced bunchgrass (*Andropogon* spp.) on Hawaii has led to a dramatic increase in the frequency and range of lightning-produced fires, further damaging native plant and animal species not adapted to burning. Although goats and sheep that once nearly extirpated the famous native silversword from Haleakala (*Argyroxiphium sandwicense*) have been largely removed, pigs, rats, and mongoose (*Herpestes auropunctatus*) introduced to kill the rats have decimated ground-nesting birds and, in the case of the pigs, spread the seeds of alien plants while further reducing many of the edible natives (Stone 1989). Fencing combined with widespread snare-trapping of pigs in both parks has recently led to significant local restoration of native forests, but this controversial practice—strongly opposed by animal rights groups—will likely have to be sustained indefinitely if it is to maintain the ecological benefits achieved.

Although both Hawaii Volcanoes and Haleakala were established as national parks principally for their physical features—volcanoes and, in the case of Haleakala, the silversword as well—their value as wild nature preserves was recognized early on. Yet the rate of human-induced biotic change in these parks is so rapid that ecological relationships among natives and recent arrivals have only begun to sort themselves out. In the meantime, "weediness" is a striking feature of many areas as a few new aliens dramatically overwhelm the locals. Ironically, perhaps, this is not entirely a novel process. About twenty-five species of plants were brought by the original Polynesian settlers over a thousand years ago, as were dogs, pigs, and rats. The large native Hawaiian birds appear to have been hunted to extinction by the new arrivals.

Park management in Hawaii has expended millions of dollars to contain the spread of the most aggressive recent aliens and exterminate new arrivals. Both the brown tree snake (*Boiga irregularis*), which devastated Guam's native birds, and the European rabbit (*Oryctalagus curriculus*), famous for its denuding abilities on numerous islands, have been intercepted in recent years. With the exceptions of goats removed from both parks, and pigs from a portion, however, the ecological turnover in favor of aliens has continued. Up to one-half of all endemic Hawaiian plants are threatened with extinction, representing a vast loss of global biological diversity. Granted, the great infusion of plant and animal species from elsewhere—including most of the ornamental tropicals admired by island visitors—has contributed to a high local species diversity on the islands. But as the islands and their national parks lose local species in favor of cosmopolitans, planetary biological diversity is eroded.

In all of these cases, human activities have led to losses of native biota and—in most places—their replacement by cosmopolitan and human-adapted species from elsewhere. This process has eroded the "distinctiveness of place" of each park and diminished its native wildness in favor of a homogeneous greenscape. To most park visitors who have little familiarity with nature, the changes are unobserved and they experience the park as "nature"—in other words, escape from industrial urban life. To varying degrees the loss of local biodiversity and wildness is reversible, but only by yet another level of aggressive human intervention—this time intentional and often quite visible to park visitors. The alternative, however, is to acquiesce and accept the progressive degeneration of pre-Columbian wild systems in the national parks and their replacement by systems ever more closely resembling what exists outside the parks. As Michael Soulé (1990) observes, the rapid human reconstruction of ecological communities is global and presages a collapse of the existing paradigms of conservation biology on which modern park management has come to depend.

The Conundrum of Climate Change

Bill McKibben, in his book *The End of Nature* (1989), posits that since humans have altered planetary processes by the addition of greenhouse gases, stratospheric-ozone-depleting gases, and ubiquitous toxins, "nature" no longer exists: all life systems reflect human influence. Even if this position is too extreme, it highlights a significant paradox in the future management of national parks as wild systems that also function as preserves of native biological diversity. Changes in climate will lead to changes in the suites of species—especially plants—adapted to a given locale. Parks are increasingly becoming ecological islands as the landscapes that surround them are converted to agriculture or development. Thus while climate change can be expected to lead to the local extirpation of species in parks, the invasions of many native "replacement" species—those adapted to the new climate—will be blocked by isolation. The intentional introduction or maintenance of native species could in some cases be used to facilitate the introduction of species that would have arrived on their own before habitat fragmentation, as well as to preserve the survival of other species that would no longer be sufficiently adapted to persist under the new climatic and ecological conditions. Such intensive management is in fact likely to be needed to preserve species of plants and animals that already are local in distribution.

To manage national parks in this way emphatically abandons the contemporary ecologically based notion of wildness. We indeed become trapped into caring for the rest of life in a transformed world. The alternative is to permit life forms to sort themselves out on their own. We could then expect rapid climatic change to lead to a significant loss of biological diversity through extinction and the advantaging of cosmopolitan species, all taking place under conditions created by humans—McKibben's end of nature. Human-induced global change (not simply climate but also such changes as habitat fragmentation and introduction of weeds and pathogens) has eliminated the possibility of treating national parks as wilderness reserves that can sustain themselves as islands in time. So the dilemma is this: Park ecosystems are changing on account of global-scale human forces. Thus there is no longer a protected wild reserve to impede local-scale management of nature. But what shall be the intent of such management? The most obvious aim would be to reverse or mitigate human-caused perturbations: ignite fires where the native fire regime has been disrupted; reintroduce extirpated species; eliminate alien species. In practice, however, these levels of biotic management are frequently too expensive, are impractical, or introduce yet more artifice. Ecological engineering, however benign, deprives park visitors (and distant supporters) of the subjective experience of "nature on its own terms." What, then, is a national park?

Natural vs. Cultural Landscapes

National parks are established not only for natural features but for cultural ones as well. For many decades, the National Park Service presumed a relatively straightforward dichotomy between "natural" parks—generally the big wildernesses—and smaller "cultural parks" featuring a historic battlefield or historically significant buildings in the East or aboriginal archaeological artifacts in the West. The increasing sophistication of park managers and changes in the law have gradually led to management for both natural and cultural resources in many national parks.

A relatively new concept in cultural resource management is the "cultural landscape." For example, the large grassy clearing known as Cade's Cove in Great Smoky Mountains National Park is not self-sustaining. It is maintained by burning, clearing, and grazing to maintain and illustrate the conditions created by early settlers in the area. Of course, most of the Great Smoky Mountains was logged and farmed until the chestnut blight and subsequent establishment of the park forced its inhabitants to relocate and most clearings were invaded by trees. There is presently a tacit understanding in the park that most of it will be permitted to return to forest—although not the forest that preceded either aboriginal or European settlement—while small areas will be artificially maintained in a condition simulating white settlement of the eighteenth and nineteenth centuries.

As research continues to reveal the many ways in which "wild" park landscapes were transformed, and may continue to be transformed, by activities of aboriginal or European people, there arises a conflict about the most appropriate management scheme for landscapes: Are they to be quasi-natural landscapes without humans, in which as many of the original ecological pieces and processes as possible will be preserved or restored, or will they reflect some moment in their cultural history: Indians, cowboys, hippie commune? Among recent and serious proposals by government and academic cultural resource specialists has been a recommendation to preserve traditional campsites in park wilderness—in direct opposition to attempts in recent decades to "naturalize" wilderness by eliminating as many traces as possible of human passage and occupation. Another proposal has been to preserve as a cultural landscape some of the vast, denuded stretches produced by many years of cattle grazing on one of the Channel Islands, as well as the cattle themselves, instead of eventually eliminating grazing and attempting to restore, to the extent possible, the pre-Columbian vegetation presently confined to small refugia.

This expanded vision of cultural resources overlaid on an otherwise human-unoccupied landscape represents an unresolved clash of visions that plays itself out daily among the resource managers of the National Park Service and other wildland management agencies. While the culturalists have effectively dispelled the myth that park lands are unaffected by the past affairs of humankind, they raise the possibility—the fearsome possibility—that there then is no wild nature in parks: parks are constructions. This does not prevent a kind of park that joins the preservation of biological diversity with the preservation of cultural artifacts. Compromise is possible. But such a place recedes ever further from wilderness.

What Are Parks For?

In North America, the wilderness parks and other designated wildernesses are the closest thing we have to markers against which we can judge the world we have invented nearly everywhere else. They represent places where our species has exercised restraint, where we have resisted the wholesale conversion of material and energy to our own purposes. Whatever the "rightness" or "wrongness" of the civilization we continue to invent, wild nature and national parks represent—however imperfectly and however dependent on our continued care—ecological anchors to our own and the planet's past.

Managing national parks from "nature out" rather than from "humankind in" may well be a fiction. The trouble with managing biocentrically is that we do not really know what constitutes ecosystem management, although we now are striving to achieve it (Agee and Johnson 1988). But it is nonetheless the most conservative approach during a period of great uncertainty. If we strive to preserve all the parts, the native ecosystem elements and processes, society has the opportunity for choices in the future. With wisdom and improved scientific understanding, we may well be able to use parks to preserve most of the parts, while they continue to provide spiritual solace for as long as our society finds value in wild nature.

References

Agee, J. K., and D. R. Johnson. "A Direction for Ecosystem Management." In J. K. Agee and D. R. Johnson (eds.), *Ecosystem Management for Parks and Wilderness*. Seattle: University of Washington Press, 1988.

Anderson, R. S. "Holocene Forest Development and Paleoclimates Within the Central Sierra Nevada, California." *Journal of Ecology* 78 (1990):470–489.

Bonnicksen, T. M., and E. C. Stone. "Managing Vegetation Within U.S. National Parks: A Policy Analysis." *Environmental Management* 6 (1982):101–102, 109–122.

Graber, D. M. "Rationalizing Management of Natural Areas in National Parks." *Bulletin of the George Wright Society* 4 (1983):48–56.

——. "Coevolution of National Park Service Fire Policy and the Role of National Parks." In J. E. Lotan, B. M. Kilgore, W. C. Fischer, and R. W. Mutch (tech. coord.), *Proceedings of the Symposium and*

Workshop on Wilderness Fire. USDA Forest Service Gen. Tech. Report INT-182. Ogden, Utah: USDA Forest Service, 1985.

Graumlich, L. J. "A 1000-Year Record of Temperature and Precipitation in the Sierra Nevada." *Quaternary Research* 39 (1993):249–255.

Kilgore, B. M., and D. Taylor. "Fire History of a Sequoia–Mixed Conifer Forest." *Ecology* 60 (1979):129–142.

Leopold, A. S., S. A. Cain, D. M. Cottam, I. N. Gabrielson, and T. L. Kimball. "Wildlife Management in the National Parks." *Transactions of the North American Wildlife and Natural Resources Conference* 28 (1963):28–45.

Lewis, H. T. *Patterns of Indian Burning in California: Ecology and Ethnohistory.* Anthropological Papers, no. 1. Ramona, Calif.: Ballena Press, 1973.

Macdonald, I. A. W., D. M. Graber, S. DeBenedetti, R. H. Groves, and E. R. Fuentes. "Introduced Species in Nature Reserves in Mediterranean-Type Climatic Regions of the World." *Biological Conservation* 44 (1988):37–66.

McKibben, Bill. *The End of Nature.* New York: Random House, 1989.

Parsons, D. J., D. M. Graber, J. K. Agee, and J. W. van Wagtendonk. "Natural Fire Management in National Parks." *Environmental Management* 10(1) (1986):21–24.

Smith, C. W. "Non-Native Plants." In C. P. Stone and D. B. Stone (eds.), *Conservation Biology in Hawai'i.* Honolulu: University of Hawaii Press for University of Hawaii Cooperative National Park Resources Studies Unit, 1989.

Soulé, M. E. "The Onslaught of Alien Species and Other Challenges in the Coming Decades." *Conservation Biology* 4 (1990):233–239.

Stone, C. P. "Non-Native Land Vertebrates." In C. P. Stone and D. B. Stone (eds.), *Conservation Biology in Hawai'i.* Honolulu: University of Hawaii Press for University of Hawaii Cooperative National Park Resources Studies Unit, 1989.

Swetnam, T. W. "Fire History and Climate Change in Giant Sequoia Groves." *Science* 262 (1993):885–889.

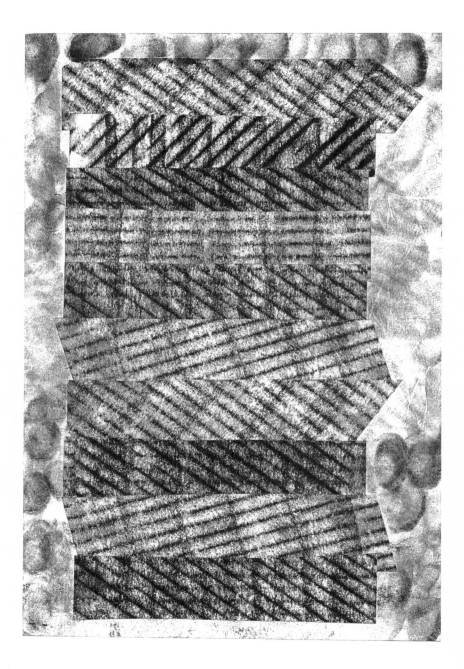

CHAPTER NINE

THE SOCIAL SIEGE
OF NATURE

MICHAEL E. SOULÉ

Living nature—the native species of plants and animals in their native settings—is under two kinds of siege; one is overt, the other covert. The overt siege is physical; it is carried out by increasing multitudes of human beings equipped and accompanied by bulldozers, chainsaws, plows, and livestock. The covert assault is ideological and therefore social; it serves to justify, where useful, the physical assault. A principal tool of the social assault is deconstruction.

Deconstruction is a style of "postmodern" critical analysis originally applied to texts. Recently deconstruction has also become popular among social critics, particularly among those who identify with emancipatory (human social justice) movements. A tenet of these movements is that one must question the premises that sustain the existing social order. And if those premises "privilege" a particular group, and if that group has not struggled to achieve its status, or if the premises are "false," then it is essential to "deconstruct" these premises—to lay them bare by the dissection of analysis—because the exposure of premises increases the likelihood of change. The reader of this book will have already learned a great deal about this.

When I discovered that social critics were deconstructing both nature and wilderness, even questioning their existence and essential reality, I wondered how and why. Gradually I have come to view many of these critics as humanists who feel that they must attack and redefine the concept of living nature and its protection as part of the struggle to liberate the less powerful classes of *Homo sapiens* from oppression by economically and politically stronger subgroups of the species.[1] For example, some argue that since people have always been part of nature, it is unreasonable to exclude native people from nature preserves.[2]

In this concluding chapter I present a status report on living nature in light of contemporary biological thought and recent global biophysical changes.[3] I also try to clarify the structure and motivation of the social assault and to deconstruct some current fallacies, misconceptions, and myths that have become impediments to protecting what remains of species diversity and wilderness. Finally, I argue that postmodern views are a major influence on the formation of biodiversity policy, sometimes to the detriment of living nature.

Coexisting Constructions of Living Nature

Euro-American civilization, the dominant culture in recent history, has invented many living natures, each formation mirroring, to some extent, the science of the day and the heterogeneity of religious beliefs and cultures. It is apparent that no universal Western perception of nature exists. People can hold many independent impressions of nature, depending on culture, experience, context, and scale. For an American, living nature can be a sunny meadow, the mountains, the coast. It can be red in tooth and claw as on television documentaries. It can be a shelterbelt or a park, a pasture or a plantation. For many, living nature is the storehouse of resources provided for human welfare. And there are people, still, who perceive everything in nature as the work of God (to teach us humility) or that of Satan (to tempt us into sin).

One of the explanations for the coexistence of such diverse views is our ignorance, both cultural and naturalistic, about our world. Many human beings are still pre-Darwinian in their understanding of natural selection (and in their faith in a wide range of fundamentalist doctrines), pre-Hellenistic in their command of logic, pre-Bernoullian in their grasp of probability, pre-Mendelian in their understanding of genetics, and pre-Modern in their humility before the dissipative significance of the Second Law of Thermodynamics. Most people, moreover, being unaware of Malthus, are oblivious to the implacability of geometric population growth in a world where resources increase arithmetically. Nor do most people appreciate the genetic/evolutionary basis of human selfishness—why reproductively competent human beings, just as in all the other ten million species, are preoccupied throughout their lives with their own biological fitness, which translates into attractiveness, wealth, and status. Thus it is ironic that we speak so glibly of postmodernism in an age when the majority of people, often because they are uneducated, remain premodern and easily fall victim to superstition and demagoguery.

It is not surprising, then, that so many concepts of living nature coexist in the so-called modern world. Perceptions of nature do not so much change over time as accumulate, layer on layer, such that the most scientific conceptions exist side by side with the most pagan, even within the mind of a single person. Figure 9.1 illustrates nine distinct cognitive formations:

- Magna Mater: the hunter-gatherer, animistic, pagan sense of divine oneness or monism.[4]
- Unpredictable and Evil Bully: a force mindlessly causing random inconvenience and catastrophes, or even a fall into anarchy and barbarism.
- Aging and Reluctant Provider: a supply depot producing fiber, minerals, food, fresh water, pharmaceuticals, and other services.
- Wild Kingdom: the venue of trophy, camcorder, and life list. This is the nature of television documentaries, of the ecotourist on safari in Kenya or Antarctica, of the grizzly fishing for salmon or mauling a camper, of the scuba-diver among coral reefs or the hiker in forest, tundra, and steppe. It is also the deer through the rifle scope and the bird through binoculars.
- Open-Air Gymnasium: the nature of ski slope, surf, rock climb, mountain ascent, and river rapids. Here the emphasis is on the sensual high of strenuous accomplishment, overcoming limits, winning contests, and domination.
- New Age Temple: the locus of vision quests, of drumming, dancing, sweat lodges, and other aboriginal-like or shamanistic rituals for facilitating rites of passage, self-esteem, or sense of

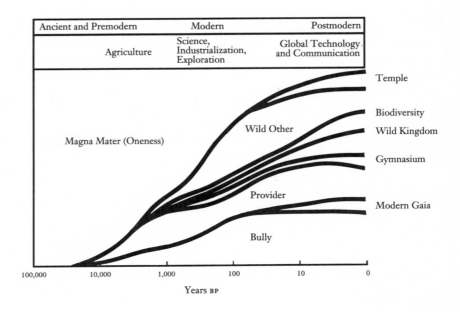

Figure 9.1 Western constructions of living nature: though the relative span of each concept is meant to suggest its prevalence at different periods in the West, these are merely guesses.

integrity and identity.[5] This is living nature as a divine setting—a temple—where one learns about the self, seeks peace, and pursues integration for psychic or psychological healing and maturation.[6] Perhaps this is as close as we modern human beings can get to the Magna Mater of pagan monism—to the world before a separate nature was invented.

- Wild Other or Divine Chaos: wild nature that has no concern for human beings except when other animals perceive us as a dangerous predator or as a possible food item.[7] To me this is the point of deep ecology—to let the other be, neither to contain it within the self nor to control it.
- Gaia: the view that living nature is homeostatic and self-regulating.[8]
- Biodiversity: the living nature of the contemporary Western biologist. It is the most concrete of all these perceptions, which makes it the most vulnerable.

This list is not meant to be comprehensive; the reader may want to add other "natures." (A similar typology—arguments for the preservation of the nonhuman world—was suggested by Warwick Fox.)[9] The point is that many living natures coexist in this late stage of modernity, a reflection of the polymorphic, fragmented nature of human occupations and preoccupations in a civilization that encompasses an extraordinary range of subcultures, levels of affluence, contact with natural habitats, and philosophical sophistication.

A Scientific View of Living Nature

What is living nature from the perspectives of late-twentieth-century taxonomists, ecologists, biogeographers, evolutionists, geneticists, and physiologists (all of whom are probably unreconstructed modernists), and how does this living nature differ from the other eight constructions?[10] Living nature can be described by reference to four categories of process: physical, evolutionary, ecological, and anthropogenic.

Physical Processes
Astronomical, geological, and climatic events are, for practical purposes, random (sometimes chaotic) and beyond control. Predictability, in most cases, is impossible or unlikely in the near future, except in the loose sense that certain events (such as earthquakes, collisions with large extraplanetary objects, major storms, and climate change, including glacial epochs) are highly probable, given enough time. Because climate and other large-scale physical events are characterized by nonlinear and often chaotic dynamics, long-term predictions may never be reliable. Though we are not in control of large-scale physical events, our activities can, nonetheless, affect many phenomena, including climate and sea level, local weather, and the ozone layer.

Evolutionary Processes
Evolutionary biology continues to reshape our view of living nature. There are five major dimensions to our evolutionary worldview. First, there is now no question that all life on earth evolved from a common ancestor. The genetic material and the codes embedded

within it reveal that every living kind of plant and animal owes its existence to a single-celled ancestor that evolved some three and a half billion years ago. All species are *kin*.

Second, the genetic material of contemporary organisms is constantly being altered by mutations, and the resulting variability is constantly pruned by differential survival and reproduction—natural selection. Natural selection is variable in intensity and often episodic. Evolutionary rates are inversely proportional to generation length, which explains why pathogens and pests evolve faster than either we humans or our crops. Most evolutionary biologists believe that natural selection has tended to increase the complexity of organisms. Complexity, however, is limited by developmental, ecological, and genetic constraints. An understanding of natural selection, as we shall see, is relevant to our current environmental and medical predicaments.

Third, the related phenomena of isolation and space account for the large number of species on the planet. Millions of species persist because mountain ranges, lakes, river systems, oceans, islands, and continents are relatively isolated; isolation facilitates speciation and space permits coexistence, in part by partitioning habitats. Organisms evolved opportunistically to take advantage of the immense range of spatial, temporal, and physical/chemical variation in habitats.

Fourth, it is probably an accident that mammals suddenly came to dominate the fauna during the last sixty-five million years. Had it not been for the doomsday scenario that apparently unfolded when one or two huge asteroids crashed into the earth around sixty-five million years ago, dinosaurs might still rule. Human beings, therefore, owe their existence to an unpredictable catastrophic event. On the scale of the solar system, however, events of this kind are both trivial and frequent.

Fifth, on the scale of human life spans, the natural extinctions of species of plants and animals are rare indeed. Although biologists have recorded hundreds of extinctions in the last few centuries, virtually all are artificial.[11] Those who argue that extinction is natural and inevitable are guilty of the "Auschwitz fallacy," the fallacy that ignores causation as well as acceleration. The important thing about contemporary extinctions is the *rate* of annihilation, not their long-term probability (which is 1). Extinction rates have increased a thousandfold in recent decades because of the ecological hegemony of *Homo sapiens*.

Ecological Processes

The real biological world little resembles the rose/blue-tinted television portrayal. Certainly the idea that species live in integrated communities is a myth.[12] So-called biotic communities, a misleading term, are constantly changing in membership. The species occurring in any particular place are rarely convivial neighbors; their coexistence in certain places is better explained by individual physiological tolerances.[13] Though in some cases the finer details of spatial distribution may be influenced by positive interspecies interactions, the much more common kinds of interactions are competition, predation, parasitism, and disease. Most interactions between individuals and species are *selfish*, not symbiotic, though these interactions can lead sometimes to evolved and reciprocally beneficial interspecies responses that may help to maintain diversity and the integrity of social groups.[14] *Homo sapiens* is no exception.[15] In general, though, the idea that living nature comprises cooperative communities replete with altruistic, mutualistic symbioses has been overstated. (See note 7.)

Moreover, living nature is not equilibrial—at least not on a scale that is relevant to the persistence of species. Nor do homeostatic systems such as Gaia buffer life on a relevant spatiotemporal scale—that of plant and animal species in real ecosystems—particularly when confronted with a species such as *Homo sapiens*.[16] In a sense, the science of ecology has been hoist on its own petard by maintaining, as many did during the middle of this century, that natural communities tend toward equilibrium. Current ecological thinking argues that nature at the level of local biotic assemblages has never been homeostatic. Therefore, any serious attempt to define the original state of a community or ecosystem leads to a logical and scientific maze. The principle of balance has been replaced with the principle of gradation—a continuum of degrees of human disturbance (see Borgmann in Chapter 3).

Anthropogenic Processes

By far the most important ecological force today is the increasing spatial and material dominance of *Homo sapiens*. First, the human population is unstable; its growth is out of control by any biological standard; the growth rate of humans is, by definition, unsustainable because it is exponential. The population of *Homo sapiens* doubles every four decades or so. But population is not the only phenomenon that is growing exponentially. The same explosive rates characterize scientific discovery, technological development, and most categories of information. In light of these rates, and given that people do not learn more today than they did one hundred years ago, we must conclude that our ignorance is also increasing exponentially.

Second, population growth, particularly when combined with powerful technologies, is causing other changes, many of which we are just beginning to detect. Nations and multinational corporations are depleting fossil fuels and topsoils, eliminating aquifers by pumping, polluting groundwater, wiping out inland seas (such as the Aral Sea), and removing nearly all ancient forests and converting them to simpler, less diverse systems. People are degrading most grasslands and destroying most of the world's fisheries, slowly replacing them with aquaculture in ponds and estuaries which destroy the nesting and breeding habitats of many species. In addition, manufacturing and consumption are perturbing the earth's atmosphere by releasing reactive chemicals—including gases that are selectively affecting terrestrial and aquatic species (such as ozone and acids)—and other chemicals that have the capacity to destroy the protective shield of ozone and increase the temperature. Among the most insidious of the active chemicals now being released are those with physiological effects. In a sense, our species is conducting an uncontrolled, planet-wide physiology experiment. For example, the falling fertility of male animals (from sturgeon to alligators, from cougars to human beings) is being attributed to estrogen-mimicking chemicals such as DDT and ethynylestradiol in birth control pills (one of the most potent biochemical agents known).[17] Together these changes are causing an episode of species extinction that is likely to eliminate one-quarter to one-half of all species during the next fifty to one hundred years.[18]

Third, as Daniel Botkin remarks, "We talk about spaceship Earth, but who is monitoring the dials and turning the knobs? No one."[19] With few exceptions,[20] the laudable objectives of natural resource sustainability are subverted by the greed of the extractors themselves. Under the Magnusson Act, for example, the fundamental fishing law in the United States, eight regional councils are authorized to set fishing limits; but these councils are dominated by the commercial fishing industry and, except in Alaska, have not been able to regulate themselves or protect the resource. Thus the industry is responsible for its own destruction.[21]

Fourth, the human condition can still be described as difficult for at least half of the world's people. In spite of the many ways that we have learned to exercise our will over water and wetlands (by damming and draining), over forests (by deforestation), and over predators and other species (by hunting, fishing, and trapping), and despite all the wounds we have inflicted on nature, many natural catastrophes still occur. Weather is still beyond the control of farmers, drivers, and vacationers. Storms still wreak havoc, as do earthquakes and wildfires. And rates of evolution among pathogens and pest species routinely outpace the rate at which technology can develop new fixes.

Attempts by affluent sections of humanity to distance themselves from weather, floods, weeds, diseases, and dangerous animals have been locally and modestly successful, though the costs of these actions (dams, loss of fisheries, loss of ancient forests, evolved resistance of pathogens to antibiotics and pesticides, and increase in cancer rates) will be borne by future generations. Many human beings are subject to chronic shortages of food, degradation of soils, and depletion of fuel and water supplies; at the same time, human populations are increasingly vulnerable to old diseases such as measles, malaria, diphtheria, and tuberculosis and to new ones such as AIDS, Lyme disease, Epstein-Barr syndrome, hepatitis C and E, Legionnaire's disease, new forms of cholera, and new viral pneumonias (such as hantavirus).

On the Cusp of Crisis

Living nature little resembles the everyday conceptions portrayed in Figure 9.1. Living nature is under siege. Though all life is related by descent from a common ancestor, one species of mammal, *Homo sapiens*, has occupied or exploited nearly every point on the planet's surface and has already co-opted much of the land's photosynthetic output.[22] Consequently, at least one-quarter of living species are on the cusp of the most profound extinction crisis in the last sixty-five million years. The welfare of living nature is further compromised by the relative insignificance of homeostatic mechanisms that operate on the ecosystem (or community) level. The metaphor of the self-regulating organism, when applied to biotic communities, is dangerously misleading. Though there are some exceptions, particularly outside the tropics, large sections of living nature will not survive the onslaught.[23]

The physical attack on living nature is frontal, converting wildlands and wild waters to farms, plantations, pastures, canals, airports, roads, highways, cities—a built, denatured, and degraded environment. Much of what remains, the remnants of native habitats, is severely fragmented, disturbed, overgrazed, overhunted, overfished, and clearcut. Increasingly, the physical attack, particularly in the tropics, is the struggle of materially subordinate members of *Homo sapiens* to survive in relatively infertile lands that they have been forced to occupy by more powerful elites and the implacable calculus of overpopulation. This brute force siege of nature, however, is not news. Perhaps more surprising is the justification of this physical siege by ideology—the social siege of nature.

The Ideological War on Nature

In this section I wish to outline the cultural or social siege of nature. This is not a conspiracy but a program that is carried out independently by traditional contestants who have distinctly different ideologies: conservative free market capitalists, humanists concerned with the emancipation and empowerment of certain social and ethnic groups, and others, including animal rights organizations. Their assault is covert because none of these groups would publicly admit to being hostile to nature or wilderness. Each group adheres to at least one of the three myths described here,[24] employing the myth that best fits its political needs of the moment. These myths are all postmodern in the sense that they reject conventional Western beliefs and biases:

- The Myth of Western Moral Inferiority—the proposition that Western attitudes toward nature are inferior to those based on other traditions (aboriginal, Eastern)
- The Myths of Constructionism—the view that nature is either unknowable (the Strong project of Hayes in Chapter 4) or that nature nowadays has been physically constituted by the economic activities of aboriginal peoples

- The Myth of the Pristine/Profane Dichotomy—the belief that the remnants of living nature have been irreversibly changed by human disturbance, so that nature today is an artificial production and therefore profane

As we shall see in the following sections, these myths have been used to justify various political agendas.

The Myth of Western Moral Inferiority

Stephen Kellert's research (Chapter 7) undermines the myth of Western moral inferiority, a canonical doctrine of New Age postmodernists who argue that many traditional Western and American attitudes reflect environmental insensitivity and are inferior to Eastern or aboriginal attitudes, which stress respect and harmony. Kellert's conclusions undermine this bit of conventional wisdom. This is not to say that Western society is environmentally virtuous. Far from it. But Western culture may not be more evil with respect to living nature than some other human traditions.

The national park movement started in the United States. There are many reasons why the modern search for ethical and practical models of conservation might have begun there. Americans, unlike Europeans or Asians, saw the widespread destruction of favorite places in their own lifetime. This horror of desecration, as well as the grandeur and wildness that remained, may have produced many of America's and the world's greatest conservationists, including Thoreau, Muir, Leopold, Carson, Brower, Ehrlich, Foreman, and Snyder.

A corollary of the myth of Western moral inferiority is the belief (an aspect of multiculturalism) that native peoples, particularly Native Americans, always behave in an exemplary way toward living nature, holding it in great reverence. Gary Paul Nabhan (Chapter 6) notes that there may have been exceptions. Further, it is likely that non-Westernized native peoples had a relatively light footprint because of their low population densities and their lack of access to livestock, firearms, and fossil-fueled vehicles and machines.[25] It also appears that aboriginal or native peoples have no special restraint when introduced to new technologies.[26]

The intent of those who promulgate the myth of Western inferiority is honorable—it is to discover better models for the relations of people to living nature. And many such models exist in Oriental and shamanistic traditions. The effect, however, is harmful because the myth misdiagnoses the cause of human callousness, blaming it on culture. But nearly everyone, regardless of ethnicity, wants a more prosperous and secure life. The problem is greed, whether we call it "fitness maximization" or "enlightened self-interest." And greed, as both the world's great religious traditions and evolutionary biologists agree, is more fundamental than culture, and the consequences, especially where technology-based affluence is achievable, are predictable.

The Myths of Constructionism

The deconstructionist critique of nature occurs at two levels. At the cognitive/cultural level both the existence of nature and the accuracy of descriptions of it are questioned on the ground that perceptual and cultural filters distort reality so much that we cannot have certain knowledge about it. All we have are biased reports, constructions. At the physical level, constructionists claim that nature is no longer natural. Rather, it is a human artifact because aboriginal peoples have altered it fundamentally with their economic manipulations.

Several contributors to this volume associate two objectives with the deconstructionist project—one is political. Paul Shepard (Chapter 2) describes the deconstructionist mission of Lyotard to be the leveling of all points of view: all are words (text, narrative); statements are the only reality; nature itself is inscrutable. Therefore all social theories and isms are equivalent, and "life is a struggle for verbal authority." This struggle is often viewed as the contemporary form of the Marxist class struggle of the nineteenth and twentieth centuries. The usual targets of these critiques are values and programs characterized as Western, capitalist, patriarchal, and scientific/technological in approach.

The second objective (a major project of deconstruction) is the dethroning of objectivism (Hayles in Chapter 4).[27] Objectivism, like dualism, creates a separation between the observer of nature and nature itself; it is the basis for science and the primary mode of discourse in the modern era. Objectivist biologists say that living nature, species and associations, are real—"out there"—and that science is a way of gradually increasing our knowledge of them. The deconstructionist alternative, nihilistic monism, is to deny that nature is real—or to insist that if there is anything "out there," we cannot know it because we are shut up in the concentric prisons of cultural bias and sensory apparatus. Therefore, it is impossible to know nature at all. All we have are culturally tainted reports, texts or words, including scientific studies about the world, none of which is more valid than any other. The social objective of this movement is to demystify and dethrone the "hegemonic dominance" of science and replace it in the public's ranking of authority with a level field that does not privilege any single approach or give it the "power to ignore competing representations made from other positions" (Hayles in Chapter 4).

This old scholarly trend—to ennoble words but to deny or demean nature—can be traced to ancient Greece. David Abram, commenting on Plato's *Phaedrus*, notes that the citified Socrates is ambivalent about the relationship of living nature and knowledge. While cautioning the reader to retain some skepticism about philosophizing, he expresses even more doubt about the utility of the animism that still dominated the world beyond the walls of Athens:

> Socrates, always eager for philosophical discourse, agrees to accompany Phaedrus into the open country where they may together consider Lysia's text and discuss its merits. It is summer: the two men walk along the Ilissus river, wade across it, then settle on the grass in the shade of a tall, spreading plane tree. Socrates compliments Phaedrus for leading them to this pleasant glen, and Phaedrus replies, with some incredulity, that Socrates seems wholly a stranger to the country, like one who has hardly ever set foot outside the city walls. It is then that Socrates explains himself: *"You must forgive me, dear friend: I'm a lover of learning, and trees and open country won't teach me anything, whereas men in town do."*[28]

Three thousand years later, historicists and some deconstructionists are taking the next step, claiming that living nature and wilderness are illusory—just some biologists' narrative, banal, ersatz, or, at a minimum, coterminous with history as criticized by Shepard (Chapter 2), Borgmann (Chapter 3), and Worster (Chapter 5).

A more moderate deconstructionist view is presented by Hayles (Chapter 4), who argues that if we begin with the premise that "we know the world because we are connected with it," then we might gain some insight. This perspective is attractive because it resonates with traditional wisdom. But there are dangers here. The main hazard may be the absence of effective checks and balances. How do we avoid falling into the snares of astrologers, creationists, and all manner of quacks who claim to have some special intimacy with the world? Should all points of view be given equal access to the media (a kind of extended Fairness Doctrine), and should all positions have equal funding from the National Endowment for the Humanities? Worster (Chapter 5) notes a similar problem with the excessive relativity of contemporary historicism.[29] When all norms are equal, Da Vinci is no better than graffiti. Or as Worster sarcastically notes: "Disneyland, by the theory of historical relativism, is as legitimate as Yellowstone. . . . Each is the product of history and therefore stands equal its opposite."

The bludgeon of multicultural relativism can also be used to defend oppression and cruelty. Ethnic fundamentalists, for example, might claim that female circumcision and other forms of institutionalized oppression of women cannot be condemned by a Westerner, because the Westerner, observing from a different cultural context, cannot possibly understand the social benefits of such practices. The inertia of traditional cultures often blocks attempts to educate women and free them from male domination. As stated by the director-general of Oxfam Europe: "Worldwide, the most powerful inhibitors to slowing population growth—the ones that are least understood by policy makers—are cultural, including people's views of God, ancestor worship, lineage, the purpose of the family, witchcraft, marriage, and polygamy."[30] These harsh comments about radical forms of postmodern criticism or social constructionism should not be construed as mere ethnocentrism. A level playing field is one thing; but a playing field without rules and referees is a free-for-all where bullies win.

Human beings are not the only ones that suffer from the new social polemics. Entire species are being driven to extinction because of superstitious cultural practices in wealthy countries—and the reluctance of activists to meet them head-on for fear of being called racist. Surely we must forthrightly criticize the use of tiger bones in Chinese herbal medicine and the use of rhinoceros horns for dagger handles by Yemeni men. Just as we are obligated to criticize practices that victimize minority populations of *Homo sapiens* (such as the Ku

Klux Klan "culture"), we must also struggle against an anthropocentrism that exterminates less powerful beings.

Why are some social critics in denial about the existence or significance of nature? Shepard (Chapter 2) suggests that deconstructionists suffer from sensory deprivation—a kind of developmental deficit of nature exposure that causes the solipsist hallucination that the world is illusory. Perhaps in a similar spirit, Hayles (Chapter 4) rejects the "strong program" of extreme constructivism, arguing that there is, indeed, a world outside our sense organs and brains.[31] Must we concede defeat to this postmodern nihilism?[32] Should we give up believing in pulsars, DNA, T-cells, moon rocks, shifting crustal plates, and Michael Jackson? Must all these "ideas" be rendered epistemologically equal to spirit guides and little green men? Hayles says no, referring to Sandra Harding's strong objectivism and to her own constrained interpretation of constructionism. Hayles proposes that if different cultures have similar renderings (isomorphisms) about some phenomenon in nature, then the phenomenon can be granted a kind of tentative reality.

But once the door is opened to shared viewpoints as a criterion for reality, the possibility exists that such isomorphisms give aid and comfort to those, namely Western positivist scientists, who seek generality and believe they are able to discover truths about nature. What if the realm of agreement is immense and the realm of competing constructions is small? This would reestablish a constrained realism and constitute a serious challenge to more radical constructionists, many of whom critique the "totalizing" and domineering tendencies of Western, science-dominated culture.[33] For example, if Western scientific descriptions of biodiversity agree with those of non-Western cultures, biologists might feel justified in remaining ignorant of modern philosophy and social criticism. Say, for instance, that the classifications of plants and animals developed by hunting-gathering peoples were basically the same as those developed by Western scientists. This might suggest that the differences between "Western" and "non-Western" cultural constructions of nature have been exaggerated and that the lenses of different cultures do not bend reality as much as some believe.

In fact, the taxonomies of aboriginal societies are virtually always the same in structure as those of modern, scientific cultures (both are hierarchical and consist of nested sets of exclusive categories); moreover, aboriginal taxonomies typically recognize the same entities as species as do modern taxonomists.[34] The parallel structure of Western and aboriginal descriptions of nature is referred to by E. O. Wilson:

> In 1928 the great ornithologist Ernst Mayr traveled as a young man to the remote Arfak Mountains of New Guinea to make the first thorough collection of birds, including hawks. Before departing, he visited key bird collections already deposited in European museums. By studying specimens gathered from western New Guinea, he estimated that a little more than one hundred bird species would be likely to occur in the Arfak Mountains. His species concept was of a European scientist looking at dead birds, who then sorts the specimens in piles according to their anatomy, as a bankteller stacks nickels, dimes, and quarters. Once settled in a camp, after a long and hazardous trek, Mayr hired native hunters to help collect all the birds of the region. As the hunters brought in each specimen, he recorded the name they used in their own classification. In the end he found that the Arfak people recognized 136 bird species, no more, no less, and that their species matched almost perfectly those distinguished by the European museum biologists.[35]

There are exceptions, though, to the rule that cultures have virtually identical descriptions of nature. Nabhan (personal communication) notes that both Navajo and Hopi cultures have hierarchical taxonomies, and these are quite similar at the level of genera. (Both cultures, for example, recognize pinyon pine and ponderosa pine as closely related.) Nabhan points out, however, that these cultures differ in how they classify varieties of domesticated plants.

The same is true in "Western" cultures. In fact, intracultural differences in the intricacy and accuracy of classifications are likely to be much greater than intercultural differences. This is because taxonomic discrimination is often based on need or intensity of use. Most educated inner-city Americans could not name five species of native American plants or birds, and they would not know that blue jays and ravens are more closely related than starlings and blackbirds. Nor could they discriminate between garter snakes (*Thamnophis*) and rattlesnakes (*Crotalus*). The point is that one's description of the natural world is determined less by culture than by need, but where the need is great (hunter-gatherers, ecologists, taxonomists), the constructions are quite similar. This point too was made by E. O. Wilson:

> Many years later, when I was twenty-five years old, . . . I made a long trek through the Saruwaget Mountains of northeastern New Guinea to collect ants. I repeated the cross-cultural test and found that the Saruwaget people could not tell one ant from another. An ant was an ant was an ant. This should come as no surprise. It was not that Saruwaget ants and natives failed the test, only that Papuans have no practical need to classify ants. The Arfak people are hunters who use their knowledge of bird diversity to make a living, just as European ornithologists do. In Mayr's time at least, wild birds were their principal source of meat.[36]

Aboriginal hunter-foragers need to know about the pharmaceutical and nutritional qualities of local plants. Such knowledge, however, accumulates slowly over centuries and millennia. Recent colonists, especially those who obtain their food in other ways, such as farmers and pastoralists, may have less need for a precise natural history and probably acquire it more slowly. The Navajo are relatively recent colonists to the Southwest, having arrived about the same time as the Spanish conquistadors. As they were new to the desert Southwest and soon relied on sheepherding, it would not be surprising if their knowledge of natural history were less than that of the Hopi. But the important point, I think, is that detailed aboriginal or folk taxonomies recognize the same species as do those of Western biologists. To me this suggests that cultural determinism is less important than the structure of the human sensory/perceptual apparatus, though this conclusion rests, to some extent, on the debatable premise that science is becoming extracultural (universal).

This idea—that the views from distinct cultures about the biophysical world converge—might surprise some antiscience, anti-Western social critics. Bruno Latour, for example, demonstrates that scientists are subject to fashion, bias, and ambition just like the rest of humanity.[37] Indeed, the frequent exploitation of scientists by industrialists, militarists, and other elites also undermines any claim of objectivity and neutrality. Thus it is tempting to conclude that the scientific enterprise is biased and value-ridden and that, as a project that seeks knowledge, it is no better than astrology. Though these conclusions do not follow from their premises, it is understandable that deconstructionists would argue that scientific texts are just stories, no better or worse than novels, no more authoritative than religious canons or science fiction.

But such a critique of modern science is shallow. It is easy to confuse the behavior of individual scientists with the behavior of the institution of science. Science, as an institution, is self-corrective. Science episodically but ultimately undermines the interests and even the beliefs of its own adherents. Thus the postmodern premise that individuals cannot escape from their values or from their expectations about reality is fair, but it sticks only to scientists, not science. Stephen Jay Gould has written recently:

> The factual correction of error may be the most sublime event in intellectual life, the ultimate sign of our necessary obedience to a larger reality and our inability to construct the world according to our desires. For science, in particular, factual correction holds a specially revered place for two reasons: first, because we define the enterprise as learning more and more about an external reality; second, because we know in our hearts that we can be as stubborn and resistant to change as petty bureaucrats and fundamentalist preachers—and undeniable factual correction therefore becomes a kind of salvation from our own emotional transgressions against a shared ideal.[38]

Therefore, I conclude that the nihilism and relativism of radically constructionist critiques of science and the materiality of nature, while popular in some academic circles, is sophomoric. Further, it is harmful because, as we shall see, it undermines efforts to save wildness and biodiversity. Nevertheless, the postmodern insistence on being attentive to bias and privilege, as well as its current emphasis on communication, contingency, and solidarity, are constructive, but these same elements should be applied to all species. The other constructionist myth—that nature today is not natural because it has been produced or fundamentally transformed by the economic activities of native peoples—might also be considered a corollary of the Third Myth, and is discussed in the following section.

Virginity Lost: The Myth of Pristine Nature
In one way or another, many of the contributors to this book ask us to consider and reject the following false syllogism: True nature—wilderness—is pristine; there are no completely pristine or virgin places

left; therefore true nature or wilderness is illusory; it exists no longer. If this syllogism were true, then all so-called wildlands must be artificial (Graber in Chapter 8). And if they are artificial, physically constructed, one can make the facile argument that they have no value as "nature" and so, like "subjects of rape . . . are reduced to the role of passive victims" (Nabhan in Chapter 6). Once having made this fantastic leap of logic, the ideologists license themselves and their benefactors to exploit or abuse living nature, either for profit or for human welfare, depending on which end of the political/economic spectrum they represent.

But as many authors in this volume argue, the virgin metaphor is inappropriate because virginity, like pregnancy, knows no degrees. Living nature, to the contrary, is nothing but inconstant; it knows nothing but degrees. Variability is nature's other name. Worster (Chapter 5) notes that living nature changes from one place to the next, from one moment to the next, and that change (ecological and historical) comes in many forms: cyclical, linear, fast, slow, modest, catastrophic. He then says: "But loose talk about all change being 'natural,' while true, is meaningless. We must pick and choose, consistent with ethical reasoning. . . . We cannot know which changes are vital and which are deadly." The idea that all changes are equivalent, that the loss of virginity is absolute, is a dangerous oversimplification. Borgmann (Chapter 3) makes the identical point in his "real vs. hyperreal" argument. The deceptiveness of the virgin metaphor can cause egregious mischief. As I write, at least three very distinct political movements are attempting to bolster their claims of moral superiority with this false metaphor and dichotomy: the Wise Use movement, the Animal Rights movement, and the Social Ecology and Justice movement.

The Wise Use movement, a loose consortium of conservative, anticonservation, commercial interests (real estate, cattle grazing, mining, logging, motorcycle and ORV manufacturers), argues that disturbance is not only pervasive and natural, but that disturbances caused by clearcutting, grazing, and off-road vehicles mimic natural processes and are good for ecosystems. Some of the Wise Use literature contends that because the West is no longer pristine, there should be no regulatory constraint on the pursuit of maximal short-term profits from public lands, including clearcutting and resort development in national parks and wilderness areas. Armed with such illogic, they defend government subsidies of cattle grazing on public lands and are quite successful at arousing public and political support, particularly in the western states.[39]

Essentially the same false dichotomy is exploited in a tract distributed by People for the Ethical Treatment of Animals (PETA) in 1993. It contains the following argument: The Nature Conservancy and other agencies should stop their efforts to snare pigs in Hawaii; even if the pigs are destroying the last vestiges of native forests where the few remaining species of native birds and snails still persist in remote, mountainous regions of several islands, Hawaii is not natural anymore because of the introduction of many species of plants and animals by Polynesian and European colonists; and because it is no longer natural, the native species should not be given greater weight than the survival of the pigs, who are suffering in snares, which, incidentally, are set only in remote areas where hunting is virtually impossible. (The pigs, by the way, are descendants of hybrids between large, aggressive European domesticated swine and smaller, more docile Polynesian pigs.) Thus by claiming that Hawaii is not part of nature anymore, PETA feels justified in giving greater ethical weight to the suffering of individual mammals than to the survival of entire, endemic species.

Some social ecologists use the "fallen virgin" metaphor, arguing that nature has been physically produced, and that nature is, therefore, an invention of culture:[40]

> Travelers [in the Amazon] today, as over the last four centuries, believe they are observing "natural" forest, but this forest is most likely the product of human decisions of the past and even today. . . . Wherever one turns, the landscape almost invariably bears the imprint of human agency, starting with fire.[41]

The authors then go on to claim that this fallen state legitimates further humanization and exploitation, even by people who have had no cultural contact with the forest until this century. The basis for their theory of human construction is that the Amazon forests, as well as forests in other parts of the tropics, have been perturbed by human activities for thousands of years, as evidenced by charcoal in the soil and by the widespread abundance of native fruit and nut trees. The conclusion of human biophysical "production" of the forest by Amerindians is then used to justify further material refashioning.[42]

Is it true that wildlands have been materially fashioned by the economic activities of humans? And if so, does this mean that living nature is produced or constructed by people? Nabhan (Chapter 6) and Graber (Chapter 8) agree that aboriginal groups have indeed altered ecosystems by the use of fire, selective harvesting, selective plantings, and similar economic activities. Nabhan notes that the American biota and landscape were significantly altered by fire before the arrival of Europeans. Moreover, the ancestors of the American Indians probably exterminated most of the native megafauna (75 percent of the genera of large mammals, including mastodons, ground sloths, camelids, equids, and indirectly their large predators) of the New World about ten thousand years ago.[43] Such depredations have no doubt led to major changes in the relative abundances and geographic distributions of these native species. Nabhan also mentions the case of palms that have been planted outside their original ranges.

Can it be claimed, though, that such changes in the absolute and relative abundances of species constitute the invention, fashioning, or construction of nature? The answer is a matter of definition. I might claim, for example, to have invented a new language because I use certain words (such as "invent" or "species") more frequently than other writers. By this artifice, every utterance is an invention of language—an utterly trivial truism. Similarly, if all the species living in a region evolved in place (or on that continent or island) and exist nowhere else, then quantitative changes in the distribution and abundance of some economically important plant species hardly constitute an "invention."

To claim that *Homo sapiens* has produced or invented the forest ignores the basic taxonomic integrity of biogeographic units: species today still have geographic distributions determined largely by ecological tolerances and geological history and climate, rather than by human activities. With some exceptions, such as the higher latitudes in the Northern Hemisphere, geological and other biogeographic influences have led to the evolution of distinctive continental biotas, particularly in tropical, southern, and oceanic regions. This is because Africa, South America, South Asia, and Australia have been isolated by climate or oceans for more than one hundred million years. For these reasons, the floras and faunas of subtropical and tropical regions on different continents contain less than 1 percent of their species in common, excluding commercially exploited zones and towns.

The species that occur in most seminatural places (except islands) are predominantly species that evolved in situ. During the last five thousand to forty thousand years, however, depending on the continent, human economic activities, particularly hunting, farming, pastoralism, logging, irrigation, and urbanization, have altered the distributions and abundances of some species and have caused the extinctions of larger prey species and their predators. Moreover, humans have introduced species including thousands of kinds of plants plus rats, mongooses, house cats, pigs, sheep, goats, starlings, predatory snails, toads, earthworms, and fish, among others. Many of these species have become naturalized—particularly in disturbed situations (such as cultivated lands) and on islands, where they have caused the extinction of many endemic species (species occurring nowhere else). It is noteworthy, though, that cases of successful colonization in relatively undisturbed, closed-canopy wildlands on continents are exceptional.[44] It is bending the language, therefore, to state that human beings have physically constructed the wild or semiwild parts of living nature, even if the intention behind this conceptual construction is humanistic.

A corollary, I think, of the fallacy of human construction of living nature, at least the wilder parts, is the politically correct assertion that local peoples are always the best managers. According to Hecht and Cockburn: "Those who have made the forest their home are also its most accomplished masters."[45] Such a statement may require qualification. Much depends on how long they have been in the region, how mechanized they are, and their population density. Local peoples must be participants in conservation, at least where they have a stake in the resources or where their hostility will doom the project. But long-term tenure does not necessarily guarantee a benign attitude toward biodiversity or even toward the natural resources on which long-term economic welfare depends. Witness the destruction of soils, old-growth forests, and fisheries in the United States, often by local people who have been exploiting the resource for many generations.

In summary, then, the myths of Western moral inferiority, construction, and ecosystem virginity are not without policy implications. These myths are used to justify human hegemony, expansionism, and the conversion of wildlands to human uses. If policymakers were to accept these postmodern myths they might become even more accommodating to invasive or destructive practices, believing that "nature isn't natural anyway." Conservationists should wean the public from the sacred grail of Pristine Nature while warning of the opposite extreme: the profane grail of Sustainable Development—the odd delusion of having your cake and eating it too. In most cases, *Homo sapiens* destroys the very resources on which it depends, especially under conditions of rapid population growth and technological/social change. And one or both of these conditions prevail virtually everywhere.

Management Implications

Why this book? Gary Lease and I were concerned that the wave of relativistic anthropocentrism now sweeping the humanities and the social sciences might have consequences for how policymakers and technocrats view and manage the remnants of biodiversity and the remaining fragments of wilderness. The foregoing chapters do not assuage our anxiety. It is apparent that the myths of postmodernism are politically potent, and to treat them as if they were merely quaint, academic curiosities would be a mistake. The following section outlines some of the policy and management implications of the postmodern myths.

Should We Actively Manage Wildlands and Wild Waters?
The decision has already been made in most places. Some of the ecological myths discussed here contain, either explicitly or implicitly, the idea that nature is self-regulating and capable of caring for itself. This notion leads to the theory of management known as benign neglect—nature will do fine, thank you, if human beings just leave it alone. Indeed, a century ago, a hands-off policy was the best policy. Now it is not. Given nature's current fragmented and stressed condition, neglect will result in an accelerating spiral of deterioration. Once people create large gaps in forests, isolate and disturb habitats, pollute, overexploit, and introduce species from other continents, the viability of many ecosystems and native species is compromised, resiliency dissipates, and diversity can collapse. When artificial disturbance reaches a certain threshold, even small changes can produce large effects, and these will be compounded by climate change.[46] For

example, a storm that would be considered normal and beneficial may, following widespread clearcutting, cause disastrous blow-downs, landslides, and erosion. If global warming occurs, tropical storms are predicted to have greater force than now.

Homeostasis, balance, and Gaia are dangerous models when applied at the wrong spatial and temporal scales. Even fifty years ago, neglect might have been the best medicine, but that was a world with a lot more big, unhumanized, connected spaces, a world with one-third the number of people, and a world largely unaffected by chain saws, bulldozers, pesticides, and exotic, weedy species.

The alternative to neglect is active caring—in today's parlance, an affirmative approach to wildlands: to maintain and restore them, to become stewards, accepting all the domineering baggage that word carries. Until humans are able to control their numbers and their technologies, management is the only viable alternative to massive attrition of living nature. But management activities are variable in intensity, something that antimanagement purists ignore. In general, the greater the disturbance and the smaller the habitat remnant, the more intense the management must be. So if we must manage, where do we look for ethical guidance?

Where Are the Oracles?

Conceptual constructions of wild nature differ greatly among world cultures, and there is a tendency in the West to consider the Judeo-Christian-Baconian culture, as it pertains to wild nature, to be mor-ally inferior to non-Western traditions (Kellert in Chapter 7). This myth has led guilt-ridden Westerners to glorify the environmental ethics of non-Western traditions. Thus environmentalists (and a few conservationists) hold up Oriental, aboriginal, and Third World per-spectives as models for emulation.

In the same vein, the belief that Third World and Fourth World (aboriginal) peoples manage wildlands with more grace and sympa-thy than do dualistic Westerners needs to be carefully examined. As David Johns has pointed out,[47] the benign treatment of wildlands that is often associated with tribal or clan societies depends on two fac-tors: a low population density (high mortality or birth control) and their limited exposure to consumerism and modern technology. Thus it is risky to conclude that non-Western peoples are always the best stewards of relatively undenatured wildlands, particularly when they are being integrated into the growth- and consumption-oriented mainstream. Policymakers should look carefully when any group makes claims of ethical superiority.

To read this as vilification of peasants would be wrong. Nevertheless, both the elite owners and managers of large estates and the poor refugee-settlers who are forced to farm and ranch on the marginal soils of tropical forests often degrade these lands and wipe out their biodiversity. The fact that the settlers are typically victims of economic discrimination, government ineptness, and corruption does not bear on whether the poor African or South American farmer is a good manager of wild nature and natural resources. Some indigenous peoples can provide excellent guidance, some not. In most cases, the best policy is a mix of science, economics, anthropology, sociology, and local native knowledge, if it still exists.

Who Really Makes Policy?
Management implies the imposition of human values on living nature. We must ask: Whose values are being implemented? Has the social siege made inroads into policy? Adherents of the Wise Use movement certainly have a great deal of access to powerful politicians. Anyone who has followed the odyssey of many environmental bills in the United States Congress knows this. Further, no one in the biomedical field doubts the power of the Animal Rights movement.

What about the power of the Social Justice movement? Today's management policies for wildlands, particularly at the level of international conservation agencies and lending institutions, are guided more by postmodern humanism than by conservation biology. Biologists rarely make biodiversity policy. Conservation policy is made by bureaucrats, technocrats, planners, development specialists, lawyers, and economists. Their views often determine how governments decide to manage wildlands and biodiversity, or if they should be managed at all. These professionals are employed by governments, by international development agencies, by large environmental and conservation organizations, and by the World Bank. They are trained by professors in the humanities and social sciences, many of whom are sympathetic to constructionist views. The best of these professionals bring to their work a solidarity with the poor who are struggling against systemic corruption, North/South inequities, and injustice, and they remind us all of the dangers of ethnocentrism, bias, and the myths of Western superiority that nourish them. But these sentiments do not suffice.

These same professionals were trained by people who may be uncritical about claims of the moral inferiority of Western concepts of biodiversity protection, who may accept outmoded myths of ecology, including balance and homeostasis, who may be uncritical about the myth of pristine nature, who often distrust or fear genetics, who are usually uncomfortable discussing human population stabilization and optima, who may believe that sustainable development is possible in a context of economic and population growth, and who are often sympathetic to the view that scientific evidence is no different than literary criticism.

Finally, most politicians and bureaucrats are city people. The influence of city people will increase as the world becomes more urban. This is one of the quietest and most profound changes of consciousness that has occurred in the twentieth century. It does not portend well for informed, compassionate decisions about the future of wild nature.[48]

Notes

1. I refer to human beings as *Homo sapiens* because I wish to emphasize that our species is just one among millions of species of animals, although we are the most powerful animal to have evolved here, in terms of both number and destructive capability.

2. J. S. Adams and T. O. McShane, *The Myth of Wild Africa* (New York: Norton, 1992), p. 239; S. Hecht and A. Cockburn, *The Fate of the Forest: Developers, Destroyers and Defenders of the Amazon* (New York: Harper Perennial, 1990).

3. By living nature I mean biodiversity: organisms and their interactions, including nonliving constructions (nests, mounds, burrows, reefs, cavities, dams) and the conditions they create. This idea differs from the definition of "biosphere" only in that the emphasis is on living organisms, not the three-dimensional regions they occupy. "Wild nature" or wilderness is that section of living nature that is relatively unperturbed and retains a sense of solitude and independence; it is a subset of living nature. See D. Foreman and H. Wolkie,

The Big Outside: A Descriptive Inventory of the Big Wilderness Areas of the United States (New York: Harmony Books, 1992).

4. See M. Oelschlager, *The Idea of Wilderness* (New Haven and London: Yale University Press, 1991).

5. I do not wish to sound pejorative; these New Age activities have important personal and social functions. This construct of nature resembles, I think, the synergetic cosmic wilderness of Oelschlager (chap. 10).

6. Robert Bly, *Iron John: A Book About Men* (Reading, Mass., and New York: Addison-Wesley, 1990); Warwick Fox, *Toward a Transpersonal Ecology: Developing New Foundations for Environmentalism* (Boston: Shambhala, 1991).

7. For "wild nature" see G. Snyder, *Turtle Island* (New York: New Directions, 1969); see also N. Evernden, *The Social Creation of Nature* (Baltimore and London: Johns Hopkins University Press, 1992). Evernden's reference (p. 118) to "ultrahuman" is similar.

8. J. S. Lovelock, *Gaia: A New Look at Life on Earth* (Oxford: Oxford University Press, 1979). Notwithstanding evidence for a degree of atmospheric feedback or homeostasis, this idea, as discussed in the following section, is generally inapplicable at spatial and temporal scales relevant to the distribution and survival of plant or animal species. It therefore leads to a false sense of security.

9. Fox, *Toward a Transpersonal Ecology*, p. 160; see also Roderick Nash, *Wilderness and the American Mind*, 3rd ed. (New Haven: Yale University Press, 1992).

10. I do not incorporate, except as a kind of cultural, scientific weltanschauung, the Einsteinian and related contributions of modern physics such as relativity, quantum mechanics, and uncertainty, which, in most situations, are irrelevant anyway because most processes affecting living nature occur in the middle "Newtonian" range of spatial and temporal scales.

11. See P. Ehrlich and A. Ehrlich, *Extinction: The Causes and Consequences of the Disappearance of Species* (New York: Random House, 1981); E. O. Wilson, *The Diversity of Life* (New York: Norton, 1992). For example, even a species as "extinction-prone" as the California

condor would be doing fine if it were not for people. Condors are very sensitive to lead poisoning; it is the presence of lead shot and spent bullets, plus poaching and the usurpation of much of their habitat, that has caused their near extinction.

12. This and other ecological myths have been discussed by many writers. See E. C. Pielou, *After the Ice Age: The Return of Life to Glaciated North America* (Chicago: University of Chicago Press, 1992); D. B. Botkin, *Discordant Harmonies: A New Ecology for the Twenty-First Century* (New York and Oxford: Oxford University Press, 1990); D. Ehrenfeld, *Beginning Again: People and Nature in the New Millennium* (Oxford and New York: Oxford University Press, 1993); S. L. Pimm, *The Balance of Nature?* (Chicago: University of Chicago Press, 1991); M. B. Davis, "Climatic Instability, Time Lags and Community Disequilibrium," in *Community Ecology*, edited by J. Diamond and T. J. Case (New York: Harper & Row, 1984), pp. 269–284. See also Worster (Chapter 5 in this volume).

13. For references see M. E. Soulé, "The Onslaught of Alien Species, and Other Challenges in the Coming Decades," *Conservation Biology* 4 (1990):233–239.

14. The cells of plants and animals, however, are symbioses in the sense that their organelles are descendants of microorganisms that lived independently before the evolution of eucaryotic cells.

15. Regardless of cultural context, we behave like any other species: selfishly. Even antiscience critics of Western approaches to nature conservation note that "given the chance, indigenous people will exploit their environment to their advantage, using whatever technology is available"; Adams and McShane, *The Myth of Wild Africa*, p. 239.

16. J. S. Lovelock, *The Ages of Gaia* (New York: Norton, 1988); L. Margulis, "God, Gaia, and Biophilia," in *The Biophilia Hypothesis*, edited by S. R. Kellert and E. O. Wilson (Washington, D.C.: Island Press, 1992), pp. 345–364.

17. J. Raloff, "The Gender Benders," *Science News* 145 (1994):24–27. It is thought that ethynylestradiol from birth control pills is excreted into the urine and finds its way, via toilets and sewage treatment plants, to rivers and coastal waters. A sad irony is that as some

humans attempt to regulate their own fecundity with birth control pills, the pharmaceuticals they employ may inadvertently be regulating the fecundity of the very species they are trying to help.

18. Wilson, *The Diversity of Life*, chap. 12.

19. Botkin, *Discordant Harmonies*, p. 192.

20. The International Whaling Commission has been an exception, as has the Convention on International Trade in Endangered Species (CITES). CITES, however, does not regulate habitat destruction or trade within countries.

21. Timothy Egan, "Fishing Fleet Trawling Seas That Yield Many Fewer Fish," *New York Times*, 8 March 1994, p. 1.

22. P. M. Vitousek, P. R. Ehrlich, A. A. Ehrlich, and P. A. Matson, "Human Appropriation of the Products of Photosynthesis," *BioScience* 36 (1986):368–373.

23. Nor will many aboriginal groups. They are meeting a parallel fate, but the linkage between the two is frequently overstated by spokespersons of the international conservation community. Indeed, the loss of either may exacerbate the condition of the other, but it is much easier to save biodiversity than cultural diversity. See M. Gadgil and F. Berkes, "Traditional Resource Management Systems," *Resource Management and Optimization* 8 (1991):127–141. The belief that cultural diversity and biodiversity are interdependent has become a central dogma of the sustainable development community, but it is a slim reed on which to build conservation in many parts of the world. It may work for a time in special situations where population densities are relatively low, where traditional human groups such as tribal peoples in India or Amerindians in Brazil protect certain components of biodiversity, and where political pressures to protect these ethnic minorities can be brought to bear.

24. Here "myths" means conceptual models of nature or even specific premises of natural law that may or may not have empirical support but nevertheless are culturally influential.

25. D. M. Johns, "Wilderness and Human Habitation," in *Place of the Wild: A Wildlands Anthology*, edited by D. Burks (Washington, D.C: Island Press, 1994).

26. M. E. Soulé, "Conservation: Tactics for a Constant Crisis," *Science* 253 (1991):744–750.

27. This dethroning of objectivism would create a dilemma for exploiters, including the backers of the Wise Use movement, at least according to Evernden in *The Social Creation of Nature*. He remarks that modern, materialistic biology has found no evidence for the soul; even worse, biology views human antics in purely materialistic (objective) and Darwinian (sociobiological) terms. Such audacity, according to Evernden, amounts to a "categorical heresy" undermining the axiom that human beings have free will. To reject this biological materialism creates an ontological crisis: if human beings lack a special soul (or if there are no dependable criteria—such as sentience—that distinguish us from beasts), then our privileged position (to dominate the earth) cannot be logically sustained, striking at the heart of anthropocentrism (humanism) and leaving us in a "schizoid" state, "accepting the system called 'Nature' while resisting attempts to explain ourselves by it." This so-called dilemma, according to Evernden, requires the abandonment of the old Cartesian dualism and the adoption of a new "ultrahumanistic" monism that somehow enfolds divinity, wildness, and the world (p. 95nn). But will this existential dilemma prevent Weyerhauser and MAXXAM from logging the last stands of ancient forests in North America?

28. From a forthcoming book by David Abram: *Inconceivable Earth: Animism, Language, and the Ecology of Sensory Experience* (New York: Pantheon Books), chap. 4. It may appear paradoxical that city-dwelling intelligentsia donate more money to environmental charities than do similarly educated rural people. I suspect this urban generosity represents a nostalgia for nature, a flickering biophilia that can be evoked by fund-raisers seeking monetary sacrifices to totems such as bears, elephants, wolves, whales, owls, and other mythic creatures.

29. Some philosophers believe that Jacques Derrida, one of the founders of this school, never claims that all interpretations are equal in value; they say that he is not a relativist and that his point is that a text has no final or right interpretation. But his countryman Bruno Latour castigates Derrida and his followers for "mak[ing] fun of the belief in reality" and for stating that belief in reality "would betray

enormous naiveté." See Bruno Latour, *We Have Never Been Modern* (Cambridge: Harvard University Press, 1993), p. 6. Nevertheless, social and political ecologists—for example, Carolyn Merchant, *Radical Ecology: The Search for a Livable World* (New York: Routledge, 1992)—often exaggerate the subjectivity of scientific knowledge.

30. *Earthwatch*, Nov./Dec. 1993 issue.

31. Here reverberates an ancient philosophical discourse: the trustworthiness of sensory data about the natural world. Even in pre-Socratic philosophy there was a tension between those, such as Heraclitus, who trusted the senses as honest witnesses of the ever-changing flux, though he acknowledged that facts alone are not a sufficient basis for insight. In contrast, Parmenides of Elea, anticipating Plato, suspected that the senses could not be trusted and that reality was insubstantial and the phenomenal world was unitary, eternal, and undifferentiated—in other words, illusory. Perhaps Parmenides, like Socrates, was a city-man. In Buddhist thought, too, there is the appearance of a similar tension. In this case it is between Samsara (everyday, ordinary life) and Nirvana (the absoluteness, oneness, or emptiness of reality); the two poles are resolved in the Mahayana synthesis as different facets or dimensions of the same reality. As the sutra says: "The absolute works together with the relative, like two arrows meeting in midair."

32. The Western doubt about the concreteness of the natural world (Lease in Chapter 1) was given a giant push when the Hebrew God married the Greek soul—psyche (Oelschlager in Chapter 2). The monotheistic supernaturalism of the Hebrews removed divinity from nature and placed it outside. The Socratic-Platonic belief in the eternal soul and the idea of the Divine Artisan, catalyzed by Christian conceptualization of time as the linear unfolding of history, all combined to cause a revolutionary shift of consciousness. Oelschlager (p. 66) quotes Ortega y Gasset's summary of the momentous change of perspective, for "what had seemed real—nature and ourselves as part of it—now turns out to be unreal, pure phantasmagoria; and that which has seemed unreal—our concern with the absolute or God—that is the true reality. This paradox, this complete inversion of perspective, is the basis of Christianity." See José Ortega y Gasset, *Man and Crisis*, translated by Mildred Adams (New York: Norton, 1962),

p. 135. The result was a profound and irreversible drift away from the identification of our species with the divinity of the vital world; earthly things were of no significance. "Nature" became an unimportant region within the universe, humanity outside, divinity in hominid body above (Borgmann in Chapter 3). Reality became a stuff to study and control, a collection of objects, and the process of objectification and distancing (Shepard in Chapter 2) had begun, at least, in Europe. This separation has been repeated in thousands of other places and times, often with the aid of trade, occupation, and colonization—first when Greek, then Roman, then European culture diffused through the rest of the species. This Western invention of separation is blamed (I think incorrectly) for the rapaciousness of Japanese industrial culture (Kellert in Chapter 7).

33. See, for example, D. J. Haraway, *Simians, Cyborgs, and Women: The Reinvention of Nature* (New York: Routledge, 1991). Apparently deaf to this discourse, scientists go right on digging up facts and building models of natural processes, many of which appear to work in daily life, including the computer on which this is written. This stubborn provincialism of scientists may be the reason why many nonscientists appear surprised to learn that "truth" in science is acknowledged by most mature scientists to be contingent, a matter of consensus.

34. C. H. Brown, *Language and Living Things: Uniformities in Folk Classification and Naming* (New Brunswick, N.J.: Rutgers University Press, 1984); B. Berlin, P. H. Raven, and D. Breedlove, *Principles of Tzeltal Plant Classification: An Introduction to the Botanical Ethnography of a Mayan-Speaking People of Highland Chiapas* (New York: Academic Press, 1974).

35. Wilson, *The Diversity of Life*, p. 42.

36. Ibid., p. 43.

37. B. Latour, *Science in Action: How to Follow Scientists and Engineers Through Society* (Milton Keynes, England: Open University Press, 1987).

38. S. J. Gould, *Eight Little Piggies: Reflections in Natural History* (New York: Norton, 1993), p. 452. Any experienced scientist will admit that most of his or her colleagues would almost prefer to die, clutch-

ing their ideas to their breasts, than admit error or give up a favorite theory. Recognizing this simple psychological truth, Thomas Kuhn points out that the mechanism of change in scientific thought is not gradual and evolutionary but punctuated and revolutionary; see T. Kuhn, *The Structure of Scientific Revolutions* (Chicago: University of Chicago Press, 1962). Many paradigm shifts await the accumulation of an enormous weight of anomalous results plus a new theory that accommodates these perplexing data. However, the truism that personal ambition and pride often motivate the attacks on current paradigms is not an effective critique of science, only of human nature.

39. Alston Chase argues some of these points in his syndicated columns. See *Range Magazine* (Winter 1994).

40. For an excellent description of the Social Ecology movement see R. Eckersley, *Environmentalism and Political Theory: Toward an Ecocentric Approach* (Albany: State University of New York Press, 1992).

41. Hecht and Cockburn, *The Fate of the Forest*, pp. 34–37.

42. Some commentators on the tropical rainforests of West Africa argue similarly. See Adams and McShane, *The Myth of Wild Africa*, n. 21.

43. P.S. Martin, "Catastrophic Extinction and Late Pleistocene Blitzkrieg: Two Radiocarbon Tests," in *Extinctions*, edited by M.H. Nitecki (Chicago: University of Chicago Press, 1984), pp. 153–190. See also J.M. Diamond, "Quaternary Megafaunal Extinctions: Variations on a Theme by Paginini," *Journal of Archaeological Science* 16 (1989):167–175; J.M. Diamond, "The Present, Past and Future of Human-Caused Extinction," *Philosophical Transactions of the Royal Society of London*, series B, 325 (1989):469–477; Wilson, *The Diversity of Life*, p. 249.

44. Aquatic habitats have demonstrated much less resistance. See J.D. Allan and A.S. Flecker, "Biodiversity Conservation in Running Waters," *BioScience* 43 (1993):32–43.

45. Hecht and Cockburn, *The Fate of the Forest*, pp. 238–239.

46. For examples see R.L. Peters and T.E. Lovejoy (eds.), *Global Warming and Biological Diversity* (New Haven: Yale University Press, 1992); P.M. Kareiva, J.G. Kingsolver, and R.B. Huey, *Biotic Inter-*

actions and Global Change (Sunderland, Mass.: Sinauer Associates, 1993).

47. Johns, "Wilderness and Human Habitation."

48. I thank Barbara Dean, David Johns, Gary Lease, and Daniel Press for their constructive critiques of this chapter.

About the Contributors

Albert Borgmann has an M.A. in literature and a Ph.D. in philosophy. Since 1970 he has taught at the University of Montana. He specializes in the philosophy of society and culture, with particular emphasis on technology. Among his publications are *Technology and the Character of Contemporary Life* (1984) and *Crossing the Postmodern Divide* (1992).

David M. Graber began his work in the national parks in 1974 with a study of the ecology and management of black bears in Yosemite National Park that led to a Ph.D. at the University of California, Berkeley. Since 1980 he has been a research biologist at Sequoia and Kings Canyon National Parks, first with the National Park Service and more recently with the National Biological Survey. His research has centered on the application of community ecology and species/habitat relations to conservation of wildland systems and the development of appropriate biological survey methods for those systems.

Alan Gussow, a pioneer in the art and ecology movement, has held forty-one solo exhibitions, and his work is represented in fifteen museum and public collections. A winner of the Prix de Rome in painting, he was presented with an award in art from the American Academy and Institute of Arts and Letters in 1977. Gussow's work reflects his belief in the ecological viewpoint, a vision beyond boundaries in which the role of the artist is to provide meaning and form to experiences in order that those experiences might be shared. He has written extensively on ecological issues and is the author of *A Sense of Place: The Artist and the American Land* (1972).

N. Katherine Hayles is a professor of English at the University of California, Los Angeles. She holds advanced degrees in chemistry and English and worked as a research chemist before becoming a literature professor. She has published widely on contemporary literature and science and is past president of the Society for Literature and Science. Her current research focuses on the impact of information technologies on contemporary literature and science, from World War II to the present. She is presently a participant in the Reinventing Nature seminar at the Humanities Research Center at the University of California, Irvine.

Stephen R. Kellert is a professor at Yale University School of Forestry and Environmental Studies. He has conducted extensive research and published widely on human values and perceptions relating to nature, particularly animals. He is a member of the boards of directors of the Student Conservation Association, Defenders of Wildlife, and the Xerces Society. He has received awards from the Society for Conservation Biology, the International Foundation for Environmental Conservation, and the National Wildlife Federation.

Gary Lease is a professor of history of consciousness at the University of California, Santa Cruz. He has served in a wide variety of chairships, including environmental studies for three years; since 1990 he also has been dean of humanities. He took his doctorate at the University of Munich in the history of theology. His ongoing work is concentrated in the history of religious thought in nineteenth- and twentieth-century Germany (editions of Harnack and Sohm; religion and national socialism; German Judaism; biography and study of Hans Joachim Schoeps) and late-antiquity Mediterranean religious history (religion and historiographical theories; religion and politics; religion as ideology and cultural artifact). He has published extensively in all these areas.

Gary Paul Nabhan is research director and cofounder of Native/ Seeds SEARCH, a Tucson-based conservation organization. He is a MacArthur Fellow and has been a Pew Scholar in Conservation and the Environment. His book *Gathering the Desert* won the Burroughs Medal for Nature Writing in 1986, and he is the author of five other books.

Paul Shepard combined work in art history and zoology for his Ph.D. dissertation at Yale University, which was published as *Man in the Landscape* (1967). His later books include *Nature and Madness* (1982) and *The Sacred Paw* (1992). He is Avery Professor of Human Ecology at Pitzer College and the Claremont Graduate School.

Michael E. Soulé is currently professor and chair of environmental studies at the University of California, Santa Cruz. His research interests include morphological and genetic variation in natural populations of animals, island biogeography, and conservation biology. He was the founder and first president of the Society for Conservation Biology and is a fellow of the American Association for the Advancement of Science.

Donald Worster is Hall Distinguished Professor of American History at the University of Kansas. He has published eight books on environmental history, the history of ecology, and the history of the American West. His book on the Dust Bowl of the 1930s (1979) won the Bancroft Prize in American History, and he has held fellowships from the Guggenheim Foundation, the American Council of Learned Societies, and the National Endowment for the Humanities. He was formerly president of the American Society for Environmental History and is currently director of the Program in Nature, Culture, and Technology at Kansas and advisory editor for the Cambridge University Press book series Studies in Environment and History.

INDEX

Each piece of art at chapter openings was done with pastel on paper, and in some cases black or white felt was applied.

Chapter 1, page 2, *Is nature "out there" or do we create it?*; Chapter 2, page 16, *The world of indoor people*; Chapter 3, page 30, *The real and the hyperreal*; Chapter 4, page 46, *Three views of gravity*; Chapter 5, page 64, *What is very old is likely to be wise*; Chapter 6, page 86, *What Muir called wilderness the Miwok called home*; Chapter 7, page 102, *Eastern and western*; Chapter 8, page 122, *Wilderness now functions to provide solitude and counterpoint to technological society*; Chapter 9, page 136, *Perceptions of nature don't so much change over time as accumulate, layer on layer.*